Fundamentals of Electromigration-Aware Integrated Circuit Design

Jens Lienig · Matthias Thiele

Fundamentals
of Electromigration-Aware
Integrated Circuit Design

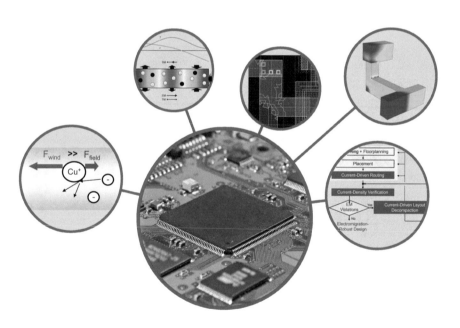

Jens Lienig
Electrical and Computer Engineering
Dresden University of Technology
Dresden, Saxony
Germany

Matthias Thiele
Electrical and Computer Engineering
Dresden University of Technology
Dresden, Saxony
Germany

ISBN 978-3-030-08811-8   ISBN 978-3-319-73558-0   (eBook)
https://doi.org/10.1007/978-3-319-73558-0

Printed on acid-free paper

This Springer imprint is published by the registered company Springer International Publishing AG part of Springer Nature
The registered company address is: Gewerbestrasse 11, 6330 Cham, Switzerland

Foreword

The stunning continued reduction in the size of IC structures is bringing interconnect reliability at the scale of individual atoms to the forefront of concern in the dependability of fielded electronic devices. That observation includes several important aspects. First is that the concern is for fielded devices, that is for voids and other material defects that come to exist and affect the component after the overall product is in operation for days, weeks, or months. Second is that the concern is for the anticipated integrity of the interconnect at the scale of individual atoms.

The physical processes responsible for these concerns are not new. These processes have always been present in fielded devices; however, the IC designs implemented by earlier technologies were robust to the defects introduced. Whereas with the present and planned IC structure scales, the material defects constitute dramatic damage or resistance change in the interconnect, thus possible failure or incorrect operation of the IC.

Specifically of concern here are migration processes: electro-, thermal, and stress migrations. The authors of this volume bring years of experience in academic and industrial circuit layout design and in the operational integrity of implemented designs to the elucidation of the current and coming design problems associated with electromigration. In particular, the authors have extensive experience in investigating devices (ICs and PCBs) that failed—often exposed to high current densities, amplified by extreme environmental conditions, for example the electronic devices that are omnipresent in automobile electronics. In this book, the authors present their knowledge for a broad audience. The information is well structured and written in a style professionals and engineering students will find accessible.

However, for both professional designers and engineering students, a full appreciation of design countermeasures for electromigration-induced defects is only possible with a deep understanding of the inner workings of electromigration. To this end, a thorough introduction to electromigration is presented and set in its relationships to thermal and stress migration processes that are also critical in this small scale. The authors augment the fundamental introduction to electromigration with the presentation of finite element modeling for analyzing electromigration effects in specific design situations.

The authors correctly argue that in the overall IC design process, the phase in which attention to electromigration mitigation should be focused is the layout-synthesis stage. In this context, present design methodology is described with options for modifying that methodology to encompass design principles for the prevention of debilitating defects due to electromigration. In addition, measures indicating the robustness of the resulting interconnect to electromigration effects are formulated from an analysis of relevant technological developments.

The extensive experience of the authors provides the basis for their advice on detailed applications for the principles in very specific and important components of ICs. Their skills as educators are evident in the presentation, as circuit designers will certainly be able to use the advice in their own designs requiring increased current-density limits with the overall goal of reducing the negative impact of electromigration on the circuit's reliability. Thus, professionals and engineering students will find it possible to apply the advice in today's IC layout design.

The continued progression of reductions in the size of IC structures expected due to developments in micro- and nanoelectronics, however, will soon require new methods. The authors, thus, turn to the future making proposals for further electromigration-aware IC design principles. These proposals are placed in the context of the future outlook in this field as a whole.

This unique book provides the fundamental science necessary for a sound grounding from which to make practical use of the complete and indispensable application-oriented information regarding the electromigration-aware design of electronic systems. It is a foundational reference for today's design professionals, as well as for the next generation of engineering students.

Charlottesville, VA, USA   Prof. Worthy Martin
                                             Associate Professor of Computer Science
                                                                     University of Virginia

Preface

Simplicity is prerequisite for reliability.

Edsger Dijkstra (1930–2002)

Today's integrated circuits are among the most complex engineering products ever built by mankind. Every day and around the world, seemingly without notice, *billions* of transistors work flawlessly in our cell phones and other electronic gadgets; the failure of a single transistor alone could render the entire system useless. Having these systems work at all is a testament to the elevated reliability of the components of which they are composed, commonly expressed in "failure in time (FIT)" units. That we define a single unit of FIT as the number of failures that occurs in 10^9 device-hours of operation, which is approximately 114,000 years, is no accident—it impressively illustrates the huge reliability requirement of today's microelectronic components.

Today's microelectronics revolution all started with the so-called first generation of modern electronics, the invention of electronic switches and miniature vacuum tubes by 1942. The first large-scale computing device, the Electronic Numerical Integrator and Computer (ENIAC), which contained 20,000 vacuum tubes, was an early and extraordinarily impressive result. However, it had reliability issues right from the start—several tubes burned out almost every day, leaving ENIAC non-functional about fifty percent of the time.

Then came the second generation, based on the discovery of the transistor in 1948. This period was mainly characterized by the switch from vacuum tubes to smaller, and much more reliable, transistors.

The 1960s saw the dawn of the third generation of electronics, ushered in by the development of integrated circuits (IC). Together with semiconductor memories, such as random-access memory (RAM) and read-only memory (ROM), they enabled increasingly complex system designs. Subsequently, we witnessed the first microprocessor in 1971. Then in 1973, Motorola developed the first prototype mobile phone, in 1976 Apple Computer introduced the *Apple I,* and in 1981 IBM introduced the *IBM PC*. These developments foreshadowed the *iPhones* and *iPads* that became ubiquitous at the turn of the twenty-first century.

As semiconductor fabrication improved, enabling larger and larger numbers of transistors to be integrated on a single chip, it became imperative for the design community to turn to computer-aided design to address the resulting problem of scale. It was an amazing self-supporting cycle: computer-aided design was facilitated by the improvements in the speed of computers, which were in turn used to create the next generation of computer chips, resulting in their own further improvement!

Throughout history, no other technical law has been as reliable and influential as Moore's Law. It fueled the PC revolution in the 1980s, the Internet in the 1990s, social media in the 2000s, and now the machine learning revolution. New electronic systems extend our senses, helping us see, helping us navigate, and helping us drive safely. Their impact reaches far beyond gadgets: electronic systems affect the way humans work and live. We have truly become a society immersed in mobile electronic devices.

Our increased dependence on electronic systems shines an intense light on their reliability. After all, a system is only as reliable as its weakest link. For example, as our chip structures become smaller and smaller, causing interconnect cross-sections to be continuously scaled down in size, we face increased migration problems, notably electromigration (EM), within our circuit interconnects. Hence, the last several years have seen a tremendous increase in electromigration-aware design approaches. It is now well-accepted wisdom that EM risks arising from ever-smaller structure sizes will become increasingly prominent in the future. If we want to continue producing working circuits in ever-decreasing sizes, we must significantly increase investment in reliability-promoting design methodologies.

This is where this book comes in. The aim is to examine the measures available for designing and manufacturing an electromigration-robust, and hence, reliable integrated circuit; to compare such measures with one another; and to investigate their use in practical, up-to-date design flows. The book not only provides a comprehensive overview of electromigration and its effects on the reliability of electronic circuits, it also introduces the physical process of electromigration and its crucial relationship with current density. The overall goal is to give the reader the requisite understanding and knowledge for adopting appropriate countermeasures.

A book of such considerable scope and depth requires the support of many. The authors wish to express their warm appreciation and thanks to all who helped produce this publication. We would like to mention in particular Martin Forrestal for his key role in writing a proper English version of our manuscript. Our warm thanks go to Dr. Mike Alexander who has greatly assisted in the preparation of the English text; his knowledge on the subject of this book has been appreciated. We thank Andreas Krinke for his contribution to analog design issues and Dr. Frank Reifegerste for the cover design. We also wish to sincerely thank Göran Jerke of Robert Bosch GmbH for his input on determining application-robust design rules, such as current-density limits (Sect. 3.4). Thanks are also due to VDI Verlag for allowing us to reprint excerpts from Matthias Thiele's PhD thesis (VDI Series 9, No. 395). We also thank Petra Jantzen of Springer for being very supportive and going beyond her call of duty to help out with our requests.

Rapid progress will continue to be made in electromigration research and electromigration-aware design in the years to come, perhaps by some of the readers of this humble book. The authors are always grateful for any comments or ideas for the future development of the topic and wish you good luck in your careers.

Dresden, Germany
Jens Lienig
Matthias Thiele

Contents

Chapter 1
Introduction

This chapter provides an overview of the evolution of microelectronics and relates it to the contents of this book, namely electromigration issues that arise during integrated circuit (IC) design and how such issues are best avoided and managed. The increasing importance of electromigration in IC design can be understood in the context of two broad developments that we explore in this chapter. First, we show that present and future development in the semiconductor industry is moving toward ever-higher current densities. And second, we discuss how boundary values for approved operation in the IC's interconnect, such as maximum tolerable current densities, are shrinking (and will continue to shrink) due to smaller structure sizes. As a consequence of these two fundamental and contradictory developments, we show how electromigration issues are becoming a crucial, and indeed in many cases critical, IC design criterion, which motivates their in-depth study in this book. The chapter concludes with an overview of the book's more detailed content, which the reader should now be able to relate to these two broad developments.

1.1 Development of Semiconductor Technology

A defining and consistent aspect throughout the history of microelectronics has been a continuous reduction in semiconductor scale, often termed "Moore's Law" [Moo65]. Year after year, we have seen circuit densities grow, as more and more transistors— each generation of transistors having smaller and smaller physical sizes—are able to be packed onto IC dies. These transistors and their interconnect are constructed of literally microscopic features, with feature resolutions of only a few tens of nanometers now being the order of the day. The trend is toward structures spanning an ever-decreasing number of atomic layers.

The primary drivers behind these reductions in semiconductor scale and the corresponding increase in density are cost efficiency and increased reliability. Small-scale semiconductors offer many benefits—for example, more functions can

© Springer International Publishing AG 2018
J. Lienig and M. Thiele, *Fundamentals of Electromigration-Aware Integrated Circuit Design*, https://doi.org/10.1007/978-3-319-73558-0_1

be integrated on an IC chip having the same surface area. In addition, the ability to perform more tasks using fewer integrated circuits also reduces costs. Alternatively, if we keep functionality the same, we can reduce chip size, which leads to lower costs and more compact systems with an increased chip count per wafer.

The desire for high reliability is one of the primary drivers for the continuous reduction in size, leveraging the specific probability that there are flaws in the wafer. In order to function properly, an IC must be located in a section of the wafer that is free of defects. Smaller chips and smaller transistor sizes increase the probability that they are located between the flaws, which increases the yield.

In addition to these benefits that arise solely from the reduced space requirements of individual transistors, there are additional drivers behind the quest for smaller structures. Small field-effect transistors (FET) have low gate capacitances, which is a favorable property as they are recharged during transistor operation. As a result, FETs can be recharged at higher frequencies with the same current, due to the reduction in the charge required.

The International Technology Roadmap for Semiconductors (ITRS) is a review of future semiconductor technologies, produced by a group of semiconductor industry experts. The detailed documents in this report provide the best guidance on the directions of semiconductor research, using time lines that extend roughly 15 years into the future. The prognoses are based on currently available technologies and extrapolations of developments that have taken place to date. It is worth noting that the report contains parameters, such as current densities and interconnect track widths, which are critical for investigating electromigration and its effects.

The technology parameters from the ITRS that are of critical interest in this book are listed in Table 1.1. As we will discuss in subsequent sections in more detail, these parameters show two alarming trends. First, while the maximum currents (upper section, last line) are shrinking, the more marked decrease in cross-sectional areas (middle section, last line) leads to increasing current densities. Second, smaller interconnects require higher current densities to perform their intended functionality (lower section, last line), while at the same time tolerable current-density limits are shrinking (lower section, upper line).

The key parameters of Table 1.1 are plotted in Fig. 1.1 to illustrate their expected trajectories.

According to ITRS predictions, the maximum possible clock frequencies for microprocessors will increase to over 8 GHz in 2024 [ITR14]. These higher clock frequencies will enable further performance enhancements for integrated circuits, including increased functionality.

Significant work will be required to reduce power dissipation, in particular to overcome thermal problems. This can be achieved by reducing currents at these increased frequencies, thereby not exhausting the increased frequency options. In addition, smaller transistors with lower voltages can be deployed as well, which further contribute to a reduction in power dissipation. For example, core voltages in CPUs with low power dissipation are as small as 0.55 V [Int17].

Table 1.1 Technology parameters based on the ITRS [ITR14]; maximum currents and current densities for copper at 105 °C

Year	2016	2018	2020	2022	2024	2026	2028
Gate length (nm)	15.34	12.78	10.65	8.87	7.39	6.16	5.13
On-chip clock frequency (GHz)	6.19	6.69	7.24	7.83	8.47	9.16	9.91
DC equivalent maximum current (μA, four gates)[a]	29.09	23.19	16.52	12.40	9.99	7.89	5.91
Metal 1 properties (Interconnect)							
Width—half-pitch (nm)	28.3	22.5	17.9	14.2	11.3	8.9	7.1
Aspect ratio	2.0	2.0	2.0	2.1	2.1	2.2	2.2
Height (nm)[a]	56.7	45.0	35.7	29.8	23.6	19.6	15.6
Cross-sectional area (nm^2)[a]	1607.2	1012.5	637.8	421.9	265.8	175.4	110.5
DC equivalent current densities (MA/cm^2)							
Maximum tolerable current density (w/o EM degradation)[b]	3.0	1.8	1.1	0.7	0.4	0.3	0.2
Maximum current density (beyond solutions are unknown)[b]	15.4	9.3	5.6	3.4	2.1	1.2	0.7
Required current density for driving four inverter gates	1.81	2.29	2.59	2.94	**3.76**	**4.50**	**5.35**
		EM to be expected			**Solutions unknown**		

Values from ITRS [ITR14]
[a]Calculated values, based on given width W, aspect ratio A/R, and current density J, calculated as follows: layer thickness $T = A/R \times W$, cross-sectional area $A = W \times T$, and current $I = J \times A$
[b]Values taken from Fig. INTC9 of ITRS [ITR14]

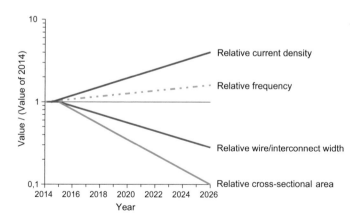

Fig. 1.1 Trajectories of key technology parameters according to [ITR14]. As current reduction is constrained by increasing frequencies, the more marked decrease in cross-sectional areas (compared to current reduction) gives rise to increased current densities in ICs. Note the logarithmic scale

1.2 Interconnect Development

The use of different materials—more especially the transitions between such materials—as well as reductions in structure size and associated technological adaptations greatly impact chip characteristics. This is especially true for the *interconnect*—the physical layout of the wires on a chip, which connect the transistors according to the network topology given by the circuit's netlist.

Copper has replaced aluminum as an interconnect material, and this has resulted in a significant change in interconnect failure modes. While copper is far more resilient to migration processes compared with aluminum, other diffusion paths are now of increased importance. In addition, copper tends to diffuse more into the surrounding dielectric. The solution to these problems required significant developments in fabrication techniques, and include new methods for patterning the metal, as well as the introduction of barrier metal layers to isolate the silicon from potentially damaging copper atoms.

The use of different types of barrier materials between copper and the dielectric is the focus of much research. It is therefore quite possible that with the advent of a new technology, barrier characteristics may change considerably. This in turn affects characteristics, such as intrinsic activation energy at the surface, the mechanical stability of the composite or process temperatures and, thus, the mechanical tension in the metallization layers.

The dielectric surrounding the interconnect likewise directly influences the interconnect characteristics. This is especially true if low-k materials, i.e., materials with a low dielectric constant k^1 relative to silicon dioxide, are used instead of the standard silica. Due to the lower stiffness of low-k materials, interconnects embedded in them are much more prone to electromigration. Indeed, the likelihood of electromigration damage is greater as mechanical loads that could prevent a possible extrusion are smaller due to the lower Young's moduli of these materials [Tho08].

Scenarios in which alternative materials are used for interconnect are described in the ITRS as well. A complete change of interconnect material has major consequences for wiring reliability. Different carbon configurations, such as carbon nanotubes (CNT) or graphene, are typical examples of interconnect materials of the future with their high current-carrying capacities (Fig. 1.2).

The choice of materials for *vias*, i.e., the vertical connections between (metal) layers, also has a tremendous impact on chip reliability and performance. In particular, the present move to three-dimensional ICs (Fig. 1.3) is raising the importance of considering via characteristics in design. The increasing deployment of

---

[1]Although the proper symbol for the dielectric constant is the Greek letter κ (kappa), such materials are referred to as being "low-k" (low-kay) rather than "low-κ" (low-kappa).

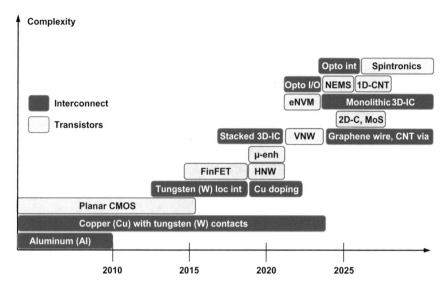

Fig. 1.2 Past and future development of interconnect and transistor technologies [YCS+13] (Tungsten (W) loc int: tungsten local interconnect (metal 0), FinFET: fin field-effect-transistor, HNW: horizontal nanowire, μ-enh: mobility enhancement, VNW: vertical nanowire, 2D-C, MoS: two-dimensional carbon (graphene), metal on semiconductor (metal gate), eNVM: embedded non-volatile memory, Opto I/O: optical input and output, NEMS: nanoelectromechanical systems, 1D-CNT: one-dimensional CNT-transistor, Opto int: optical interconnect, Spintronics: spin transport electronics)

through-silicon vias (TSVs), which are vertical electrical connections passing completely through a silicon wafer, alters the design constraints for the wiring layers [KML+12, KYL12]. On the one hand, portions of the chip layers cannot be used for routing, and on the other, there are areas in the neighborhood of TSVs where mechanical tensions impact the interconnects and transistors [PPL+11, PPP+11].

1.3   The Rise of Electromigration

Since this book focuses on electromigration (EM) issues, let us now investigate how the future developments of the semiconductor and interconnect technologies may significantly affect the electromigration problem.

As we will describe in detail in Chap. 2, the process of electromigration in the electrical interconnect of an integrated circuit (IC) is a major concern for IC designers. If electromigration is not effectively understood and mitigated during the design stage, when the IC is subsequently deployed and operated, electromigration can lead to (i) open circuits due to *voids* and to (ii) unintended electrical connections (i.e., shorts) due to *hillock* or *whisker* failures, resulting in faulty IC operation. Because of the nature of the electromigration process, which may take weeks,

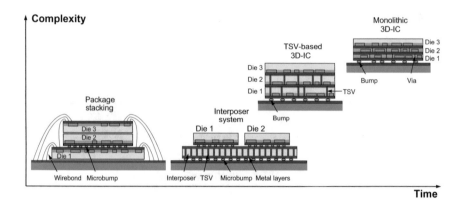

Fig. 1.3 Evolution of 3D integration technology where "vertical interconnects," such as through-silicon vias (TSVs), are gaining importance [KSE+17]. Originating with package stacking, 3D integration has evolved through interposer-based systems toward TSV-based 3D ICs and is currently on its path toward monolithic 3D ICs. While TSV-based and monolithic 3D ICs offer the highest integration densities, interposer systems facilitate an easy heterogeneous integration

months, or even years to occur, the sudden faulty operation of the IC may be particularly unexpected, detrimental, or costly (e.g., requiring a product recall).

As we will show in the following chapters, electromigration is the result of excessive current densities. The current density J is calculated from the quotient of the flowing current I and the cross-sectional area A of the interconnect:

$$J = \frac{I}{A} \qquad (1.1)$$

The development of currents in the future (which will increase I in the numerator of Eq. 1.1, thereby increasing J), and of interconnect track parameters, such as the cross-sectional area (which will decrease A in the denominator of Eq. 1.1, thereby also increasing J), are clearly critical in the context of electromigration.

By way of a first step in analysis, we note that Table 1.1 (upper section) shows a (favorable) reduction in currents over time, due to lower supply voltages and shrinking gate capacitances. However, it is important to understand that as current downscaling is constrained by increasing frequencies, the more marked decrease in cross-sectional areas (compared to current downscaling) will result in increased current densities J in ICs going forward (Fig. 1.4).

To make matters worse, the maximum tolerable current densities are shrinking at the same time due to smaller structure sizes. These concerning trends are captured in Table 1.1 (lower section). The inevitable conflict between rising current densities and falling limit values is depicted in detail in Fig. 1.5.

It is useful to analyze the trends and relationships in Fig. 1.5, to better understand the source and causes of electromigration issues. We note that the interconnect track width correlates directly with the gate length of the transistors. It equates

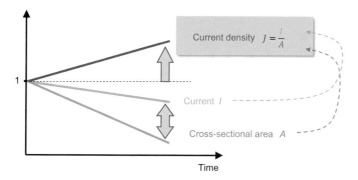

Fig. 1.4 Projected current densities over the coming years resulting from decreasing interconnect cross-sections and only slightly falling currents [ITR14, ITR16]

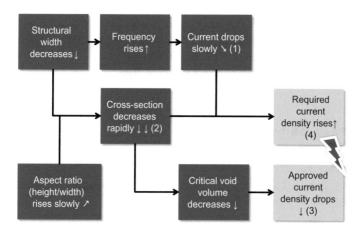

Fig. 1.5 The evolution of interconnect parameters leads to a conflict caused by the required rising current densities coupled with falling limit values (see also Table 1.1)

approximately to the smallest possible structure size of the respective technology, that is, half the grid dimension of the respective metallization layer. The on-going reduction in structure size enables higher frequencies and lower voltages. Considering the gate capacitances and associated charges needed for securely operating the transistors, the required currents in the interconnects only slowly decrease in magnitude (see (1) in Fig. 1.5). Data tells us they halve approximately every five years [ITR14, ITR16].

Furthermore, cross-sectional areas (see (2) in Fig. 1.5) are reducing in size quadratically in relation to the width within three years to about 50% of their original size. This is because the aspect ratio of the interconnects, that is, the ratio of height to width, can only be very slowly increased [ITR14, ITR16]. The effect of the down-sizing of cross-sections is that the sizes of critical voids are decreasing as well, which in turn leads to EM-induced malfunctions. Interconnect properties are increasingly subjected to side effects: while on the one hand the barrier component of the overall cross-section is rising, the specific resistance of the interconnect track is increasing due to the scattering effects of electron conduction. Both side effects accelerate the characteristic heat increase (*Joule heating*) of the interconnect. Hence, the reduction in cross-sectional area causes a reduction in allowed current densities (see (3) in Fig. 1.5) for constant durability. There is a 50% decrease in the permissible current density every three years, according to [ITR14, ITR16].

The following conclusion can also be drawn from (1) and (2) in Fig. 1.5: The current densities required to operate an integrated circuit with decreasing structure sizes double approximately every eight years (see (4) in Fig. 1.5). This is directly opposed to falling current-density boundaries (3), as these opposing trends exacerbate the problem of rising current densities. This development (4) will cause a technological hurdle even if current-density boundaries are maintained at their present levels.

If required current densities exceed approved boundaries, this will spell the death knell for technological progress as we know it in this area. And, according to the ITRS [ITR14, ITR16], approved current densities are increasingly exceeded, which makes this topic of immediate concern for IC design.

1.4   Motivation and Structure of This Book

Integrated circuits have far greater reliability than circuits consisting of discrete components; this advantage is driving semiconductor scale reductions and associated investments in advanced technologies.

Unfortunately, increasingly small IC structures begin to have a significant negative impact on reliability, as the cross-sectional areas of the metallic interconnects in the ICs are diminished in size. The problem arises because the required currents cannot be reduced to the same extent—even by reducing the supply voltages and gate capacitances. This is illustrated in Fig. 1.6, where the required current densities to drive four inverter gates, for example, increase over time as a consequence of decreased structure size.

To make matters worse, the maximum tolerable current densities are shrinking at the same time due to smaller structure sizes (see Fig. 1.5). As already mentioned, the reason for this is that small voids and other material defects, which could have been tolerated in earlier technology nodes, cause increasingly dramatic damage and side effects to the wires with shrinking metal structures. Thus, maximum tolerable current densities will have to decrease to maintain the required interconnect reliability. As a result, the ITRS indicates that all minimum-sized interconnects will be

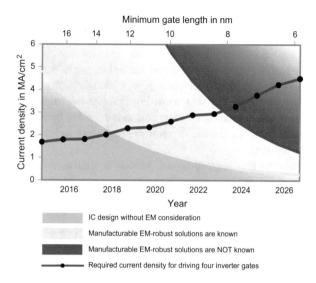

Fig. 1.6 Evolution of required and maximum current densities in IC interconnect [ITR14, ITR16]. While the required current density scales with frequency and reducing cross-section, the maximum tolerable current density is shrinking due to smaller structure sizes (cf. Fig. 1.5). EM degradation needs to be considered inside the yellow area. As of now, manufacturable solutions are not known in the red area

increasingly EM-affected, potentially limiting any further downscaling of wire sizes (Fig. 1.6, yellow barrier).

Furthermore, the total length of interconnect per IC will continue to increase. As a consequence, reliability requirements per length unit of the wires need to *increase* in order to *maintain* overall IC reliability. This accepted wisdom is contradicted by the future *decrease* in interconnect reliability due to electromigration—as noted above. The ITRS thus states that no known solutions are available for the EM-related reliability requirements that we will face in the near future (Fig. 1.6, red barrier).

Measures to handle electromigration, such as current-dependent routing or the adaptation of the track width in highly loaded interconnects, are *de rigueur* today for designing analog integrated circuits. As a result of structural miniaturization, digital integrated circuits are now also affected by the problem of increasing current densities and accompanying EM. Typical measures, such as increasing the interconnect width, common in analog circuit design, cannot be deployed in these much more complex digital circuits. Such measures would work against the reduction in structure size and prevent further scaling. New approaches are therefore required to avoid EM damage in digital circuits, as a result of falling semiconductor scale.

This book presents measures for layout design for avoiding damage caused by EM in both digital and analog ICs. We determine parameters for every measure so that the usability and suitability of a specific measure can be determined as a function of the technology used. Approved current densities can thus be increased at the critical places by means of local layout modifications. This ensures that current densities in the yellow section in Fig. 1.6 are tolerated as well. The aim essentially is to avoid exceeding approved current densities by enlarging reliability limits. This book provides the reader with the necessary knowledge to overcome such design challenges.

It is particularly important that the proposed measures be applied at the physical-design stage and especially for the routing step. The reason for this is that good interconnect routing allows the optimal utilization of measures for promoting EM robustness. Interventions at a later stage in the design process, typically involving layout modifications, are much less effective, because fewer modification options are available at this later stage. On the other hand, currents cannot be precisely specified before the layout is generated, as a physical network topology is needed to provide detailed current knowledge.

The fundamental physical problem of EM will be examined to the core in Chap. 2, as this knowledge is a requisite for adopting appropriate countermeasures. After first explaining the physical causes of EM, we introduce influencing factors arising from the specific circuit technology, the environment, and the design. We then investigate detailed EM mechanisms with regard to circuit materials, frequencies, and mechanical stresses. IC designers must be especially aware of thermal and stress migration; both are introduced and described in their interaction with EM.

Chapter 2 also outlines the principles of a migration analysis through simulation. This honors the importance of finite element modeling (using the finite element method, FEM) in EM analysis and enables the reader to develop and apply similar modeling and simulation techniques.

Chapter 3 presents options for modifying the present design methodology to encompass EM prevention. Analog and digital designs are considered separately in this context as the respective measures differ for both. Understanding that knowledge of the currents flowing in interconnects is a fundamental requisite for an EM-aware design flow, we will discuss the different types of currents encountered and show how sensible current values can be determined.

The key parameter for EM prevention is the maximum permissible boundary value of the current density in the wires. This parameter is, however, dependent on the intended use of the IC, which is why so-called mission profiles are created to determine such values. Chapter 3 describes how robust current-density boundary values (limits) can be determined, using application and reliability specifications.

Fundamental procedures for current-density verification are examined as well. Methods for eliminating problems, identified during current-density verification, by means of layout adjustment are presented. Finally, we put forward a number of approaches for increasing current-density boundary values, based on our assessment of current technological trends.

While Chap. 3 outlined options to address EM in today's physical design of electronic circuits, Chap. 4 describes in detail the EM-inhibiting effects that these options are based on. The goal of this chapter is to summarize the state of the art in EM-mitigating effects. This knowledge is presented such that a circuit designer can use it to increase current-density limits with the overall goal of reducing the negative impact of EM on the circuit's reliability. We will show how approved current densities can be increased by means of local layout modifications. Detailed application advice concludes each presented measure.

We also consider material-related options to reduce EM, such as surface passivation, and the use of EM-robust materials, such as carbon nanotubes.

In Chap. 5, we summarize our findings, make proposals for further EM-aware integrated circuit design, and present the future outlook in this field, along with expected developments in micro- and nanoelectronics.

References

[Int17]   8th Gen (S-platform) Intel® processor family datasheet, vol. 1, 2017, https://www.intel.com/content/dam/www/public/us/en/documents/datasheets/8th-gen-processor-family-s-platform-datasheet-vol-1.pdf. Last retrieved on 1 Jan 2018

[ITR14]   International Technology Roadmap for Semiconductors (ITRS), 2013 edn. (2014), http://www.itrs2.net/itrs-reports.html. Last retrieved on 1 Jan 2018

[ITR16]   International Technology Roadmap for Semiconductors (ITRS 2.0), 2015 edn. (2016), http://www.itrs2.net/itrs-reports.html. Last retrieved on 1 Jan 2018

[KML+12]   J. Knechtel, I.L. Markov, J. Lienig, et al., Multiobjective optimization of deadspace, a critical resource for 3D-IC integration, in *Proceedings of IEEE/ACM International Conference on Computer-Aided Design (ICCAD)* (2012), pp. 705–712. https://doi.org/10.1145/2429384.2429538

[KYL12]   J. Knechtel, E.F.Y. Young, J. Lienig, Planning massive interconnects in 3D chips. IEEE Trans. Comput. Aided Des. Integr. Circuits Syst. 34(11), 1808–1821 (2015), ISSN 02780070. https://doi.org/10.1109/tcad.2015.2432141

[KSE+17]   J. Knechtel, O. Sinanoglu, I. M. Elfadel, et al., Large-scale 3D chips: challenges and solutions for design automation, testing, and trustworthy integration, IPSJ Trans. on System LSI Design Methodology, **10**, 45–62 (2017). https://doi.org/10.2197/ipsjtsldm.10.45

[Moo65]   G.E. Moore, Cramming more components onto integrated circuits. Electronics **38** (8), 114–117 (1965). https://doi.org/10.1109/N-SSC.2006.4785860

[PPL+11]   J. Pak, M. Pathak, S.K. Lim, et al., Modeling of electromigration in through-silicon-via based 3D IC, in *61st IEEE Electronic Components and Technology Conference (ECTC)* (2011), pp. 1420–1427. https://doi.org/10.1109/ectc.2011.5898698

[PPP+11]   M. Pathak, J. Pak, D.Z. Pan, et al., Electromigration modeling and full-chip reliability analysis for BEOL interconnect in TSV-based 3D ICs, in *IEEE/ACM International Conference on Computer-Aided Design (ICCAD)* (2011), pp. 555–562. https://doi.org/10.1109/iccad.2011.6105385

[Tho08]   C.V. Thompson, Using line-length effects to optimize circuit-level reliability, in *15th International Symposium on the Physical and Failure Analysis of Integrated Circuits (IPFA)* (2008), pp. 1–4. https://doi.org/10.1109/ipfa.2008.4588155

[YCS+13]   G. Yeric, B. Cline, S. Sinha, et al., The past present and future of design-technology co-optimization, in *IEEE Custom Integrated Circuits Conference (CICC)* (2013), pp. 1–8. https://doi.org/10.1109/cicc.2013.6658476

Chapter 2
Fundamentals of Electromigration

Having shown in Chap. 1 that the future development of microelectronics will lead to more and more electromigration problems, let us now investigate in detail the actual low-level migration processes. A solid grounding in the physics of electromigration (EM) and its specific effects on the interconnect will give us the knowledge to establish effective mitigation methods during the design of integrated circuits (ICs).

We first explain the physical causes of EM (Sect. 2.1) and then present options to quantify the EM process (Sect. 2.2), which enable us to effectively characterize key aspects of the process and its effects. In Sect. 2.3, we introduce EM-influencing factors arising from the specific circuit technology, the environment, and the design. We then investigate detailed EM mechanisms with regard to circuit materials, frequencies, and mechanical stresses (Sect. 2.4).

Since EM is closely related to other migration processes, such as thermal and stress migration that also occur in the conductors of electronic circuits, we examine their interdependencies (Sect. 2.5). IC designers must be especially aware of thermal and stress migration; both are introduced and described in their interaction with EM.

Finally, Sect. 2.6 outlines the principles of a migration analysis through simulation. This honors the importance of finite element modeling (using the finite element method, FEM) in electromigration analysis and enables the reader to develop and apply similar modeling and simulation techniques.

2.1 Introduction

The reliability of electronic systems is a central concern for developers, which is addressed by a variety of design measures that include, among others, the choice of materials to best suit an intended use. As the structural dimensions of electronic interconnects are downscaled (Chap. 1), new factors that reduce reliability and that

© Springer International Publishing AG 2018
J. Lienig and M. Thiele, *Fundamentals of Electromigration-Aware Integrated Circuit Design*, https://doi.org/10.1007/978-3-319-73558-0_2

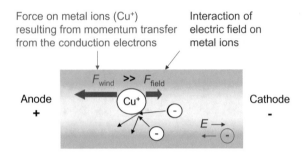

Fig. 2.1 Two forces act on metal ions (Cu) that make up the lattice of the interconnect material. Electromigration is the result of the dominant force, that is, the momentum transfer from the electrons that move in the applied electric field E

previously could be ignored now come to bear. In particular, material migration processes that occur in electrical interconnects during IC operation can no longer be ignored during IC design and development.

Material migration is a general term that describes various forced material transport processes in solid bodies. These include (1) chemical diffusion due to concentration gradients, (2) material migration caused by temperature gradients, (3) material migration caused by mechanical stress, and (4) material migration caused by an electrical field. This last case is often referred to as *electromigration*, which is the subject of this chapter (and the book); we describe its relationship to the other migration processes (1)–(3) in Sect. 2.5.

Current flow through a conductor produces two forces to which the individual metal ions[1] in the conductor are exposed, the first of which is an electrostatic force F_{field} caused by the electric field strength in the metallic interconnect. Since the positive metal ions are shielded to some extent by the negative electrons in the conductor, this force can safely be ignored in most cases. The second force F_{wind} is generated by the momentum transfer between conduction electrons and metal ions in the crystal lattice. This force, which one may visualize by analogy as a breeze or wind blowing through the leaves of a tree, acts in the direction of the current flow and is the primary cause of electromigration (Fig. 2.1).

If the resulting force in the direction of the electron wind (which also corresponds to the energy transmitted to the ions) exceeds a given trigger known as the activation energy E_a, a directed diffusion process starts. (In our earlier analogy, a leaf has been blown off the tree by the wind.) The resulting material transport takes place in the direction of the electron motion, that is, from the cathode (-) to the anode (+).

---

[1]The crystal lattice of metals is built up of ordered metal ions with an "electron fog" in-between, consisting of shared free electrons. The terms metal *atoms* and metal *ions* are considered equivalent in this context.

   The actual diffusion paths are material dependent and are mainly determined by the size of their respective activation energies. Every material has multiple, different activation energies for diffusion, namely for diffusion (i) within the crystal, (ii) along grain boundaries, and (iii) on surfaces (Sect. 2.3.1). The relationships between the individual energy levels determine which of the diffusion mechanisms (i)–(iii) dominates, as well as the composition of the entire diffusion flux.

   If one could assume the material transport was homogeneous at every location in the wiring, there would be no change throughout the interconnect: the same amount of material would be replenished as would be removed. However, the wiring of a fabricated IC chip contains numerous required features that result in inhomogeneities; as a result, the diffusion is also inhomogeneous. Among the features and resulting inhomogeneities encountered in chip designs are

- ends of interconnects,
- changes in the direction of interconnects,
- change of layers,
- varying current densities due to changes in interconnect cross-sections,
- changes to the lattice or the material,
- already existing damage or manufacturing tolerances,
- varying temperature distributions, and/or
- mechanical tension gradients.

These inhomogeneities cause divergences in the diffusion flow, leading to metal depletion or accumulation in the vicinity of such inhomogeneities. Such depletions and accumulations in turn result in damage to the interconnect, due to *voids* and interconnect breaks or *hillocks* that cause short circuits (Fig. 2.2). Another result of EM in wires is *whiskering*, which is a crystalline metallurgical phenomenon involving the spontaneous growth of tiny, filiform hairs from a metallic surface (see Fig. 2.2, right). *Whiskers* can cause short circuits and arcing in electronic circuits.

   The two types of depletion that cause damage in integrated circuits are known as *line depletion* and *via depletion* (Fig. 2.3). Electron flow from a via to a line can

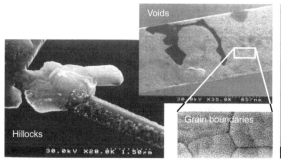

Fig. 2.2 Hillock and void formations in wires due to electromigration (*left*, photographs courtesy of G. H. Bernstein und R. Frankovic, University of Notre Dame). Whisker growth on a conductor is shown on the *right*

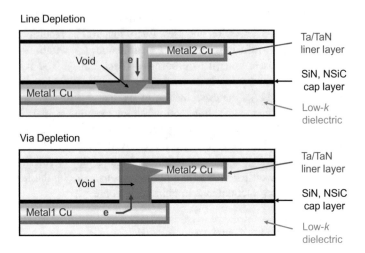

Fig. 2.3 Line depletion (*above*) and via depletion (*below*) are common failure mechanisms due to EM in integrated circuits

cause line depletion due to obstructed material flow through the cap and liner layers. Reversing the electron flow, i.e., electron flow from a line to a via, may result in via depletion, sometimes also called *via voiding*. Here too, its causes are a combination of geometry and process. As with line depletion, the material migration is hindered by the surrounding cap and liner layers. In addition, as the ratio of the line width to the via width increases, the via must carry more current for the same line current density, making the via more susceptible to the voiding process.

EM-induced damage to an IC that results from the growth of voids is further accelerated by a positive feedback loop (Fig. 2.4). Here, an initial (excessive) current density causes void growth and cross-sectional degradation, which increases the local current density. At the same time, the (increasing) current density causes a temperature rise due to (local) *Joule heating*, which occurs when an electric current passes through a conductor and produces heat. The increased heat also accelerates diffusion and thus further increases the void growth.

It is important to note that EM is only one of four different migration processes that occur in solid-state materials such as the wires on an electronic circuit. As shown in Fig. 2.5, the other processes are *chemical diffusion*, *thermal migration*, and *stress migration*, which are caused by the chemical and thermal gradients and mechanical stress, respectively. While we will consider their mutual interaction and influence on EM in Sect. 2.5, this book primarily focuses on solid-state electromigration.

In addition to the solid-state electromigration process, so-called *electrolytic electromigration* can occur in electronic circuits, often on printed circuit boards (PCBs). Its mechanisms are quite different compared to solid-state electromigration: recall that solid-state electromigration is the movement of metal within a conductive

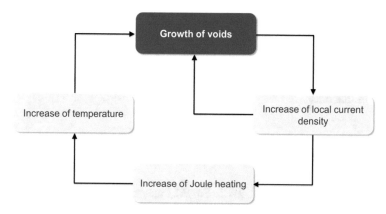

Fig. 2.4 Acceleration of the growth of voids by positive feedback at work: void growth increases current density, which in turn rises the wire's temperature due to Joule heating, which further accelerates diffusion and void growth

path due to electron momentum transfer (scattering) resulting from high current densities ($>10^4$ A/cm^2), often at higher temperatures. In contrast, electrolytic electromigration is the movement of metal across a nonconductive path at lower temperatures (<100 °C) and at low current densities ($>10^{-3}$ A/cm^2) in the presence of moisture.

Electrolytic electromigration requires moisture on the surface and a high electric field, often caused by a combination of voltage difference and narrow line spacing in a wet environment. Migrating metal ions are dissolved in an aqueous solution (e.g., water) in this process. The material flows in a direction opposite to solid-state electromigration: a DC electric field between the anode and cathode will pull the

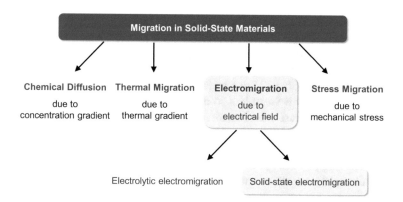

Fig. 2.5 Different migration mechanisms can occur in electronic circuits. While this book focuses on (solid-state) electromigration, their mutual interactions cannot be neglected and are covered in Sect. 2.5

free metal ions across to the cathode; hence, migration follows the direction of the electric field.

Since electrolytic electromigration can easily be avoided (by keeping the electronic circuit dry) and is visually recognizable on PCBs (tree-like structures of crystals, so-called dendrites, that traverse conductor spacing), it will not be covered further in this book.

2.2 Electromigration Quantification Options

A diffusion formula (see Sect. 2.4.1, Eq. 2.2) may be used for a detailed, quantitative analysis of the EM process, with the ultimate aim of determining the divergence in the material flow. This type of analysis yields the locations that have material accumulation and depletion, and identifies locations where initial damage is expected to occur. Such a quantitative approach also helps to obtain the mean service life duration of the interconnect.

While this analytical approach is sound from a theoretical basis, such extensive calculations are practical only in simple cases. Typically, many iterations and numerical analyses are required to determine the service life duration for even the simplest of cases. The same applies to short-term calculations for more complex architectures. As analytical solutions are often too time consuming and impractical, a numerical simulation may be necessary, even for a single change in direction or cross-section of the wire.

Fortunately, there is an empirical model for determining the *median time to failure* (MTF) for simple linear interconnects. This reliability characteristic is described by *Black's equation*, first introduced by J. R. Black in the 1960s [Bl69a], as follows:

$$\text{MTF} = \frac{A}{j^2} \cdot \exp\left(\frac{E_a}{k \cdot T}\right) \tag{2.1}$$

where A is a cross-section-dependent constant, that, among others, relates the rate of mass transport with median time to failure (MTF) [Bl69b], j is the current density, E_a is the activation energy, k is the Boltzmann constant[2], and T is the temperature.

In later variants of Eq. (2.1), the constant exponent ("2") of the current density j has been replaced by a variable n to allow the model to be applied to different types of dominant failure mechanisms. In effect, this meant different exponents were used for different interconnect materials, for example aluminum (Al) and copper (Cu). Furthermore, it has been established through studies on Al and Cu interconnects (e.g., [Hau04]) that void-growth-limited failure is characterized by $n = 1$, while void-nucleation-limited failure is best represented by $n = 2$.

---

[2]The Boltzmann constant, which is named after Ludwig Boltzmann (1844–1906), is a physical constant relating the average kinetic energy of particles with the temperature.

In the case of aluminum and its associated dominant grain-boundary diffusion, the activation energy E_a is approximately 0.7 eV for a current-density exponent of $n = 2$. Copper, by contrast, has the lowest activation energy at 0.9 eV for the dominant surface diffusion with a current-density exponent n between 1.1 and 1.3, depending on the dominant failure mode [FWB+09].

With Black's Eq. (2.1), the relation between service life duration and current and temperature can be readily estimated; the equation yields useful information for accelerated testing, as well. One caveat of the equation is that a steep rise in the current, and thus the current density, alters the failure mechanism—which is not modeled by the equation. Large temperature gradients may then arise as well, due to the characteristic heat increase of the interconnect (Joule heating), which can cause thermal migration (Sect. 2.5) or even thermal failure.

Black's equation is useful to a certain extent when designing interconnects for desired reliabilities. The main disadvantage is that the equation is designed for linear interconnects and cannot successfully be applied to entire net routes with changes in direction or changes in layer. Neither does it cover transitions between different materials and mechanical boundary conditions. This limits its usefulness, as the equation cannot therefore be used to compare different technologies. Parameters A and E_a are particularly technology-specific.

In [LT11], Li and Tan developed a different, more complex model for calculating the service life duration, which considered additional constraints, such as thermal and mechanical stress. Their model, which is based on the Eyring equation, contains far more parameters. It is easier to determine these parameters than with Black's equation, as the model is based on material properties and actual transport mechanisms. This is in contrast to Black's equation where the parameters are empirically defined and can be determined only by statistical investigations.

2.3 Design Parameters

EM-related design parameters and constraints can be divided into three groups as they are based on

- the technology, in particular the materials,
- the environment, especially the temperature, and
- the design, which is the main determinant of the current density.

We describe each of these groups in detail below.

2.3.1 Technology

The material used to construct the interconnect has a significant impact on electromigration. The key property of a conductor material is its activation energy E_a,

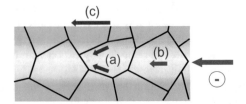

Fig. 2.6 Illustration of various diffusion processes within the lattice of an interconnect: **a** grain-boundary diffusion, **b** bulk diffusion, and **c** surface diffusion

which is a measure of the resistance of the metal ions to EM, as well as its resistance to diffusion in general. The activation energy is primarily determined by the bonding energy of the crystal metal lattice. Hence, its values are different for different interconnect materials, such as copper and aluminum. In addition, the ions in a crystal lattice have different binding energies, depending on their location within the lattice as illustrated in Fig. 2.6 and explained below.

The most stable bond, one having the maximum activation energy, is in the crystallite core. Only ions in this lattice region that are near crystallographic defects, such as voids (vacancy defects) or dislocations, are able to leave their positions. In contrast, ions at the grain boundaries in polycrystalline interconnects have weaker bonds to the lattice and thus have a lower activation energy, because the bonding forces are asymmetric. Similar behavior patterns exist at the external boundaries or surfaces of the interconnect, where the materials in the surroundings have a decisive impact on the activation energy.

In the case of aluminum, grain-boundary diffusion dominates the electromigration process, as the activation energy is lowest at the grain boundaries ($E_a \approx 0.7$ eV, Sect. 2.4.1). EM robustness therefore can be improved significantly for aluminum interconnect by doping with copper, for example, or by nucleating larger grains.

This contrasts with copper, where boundary or surface diffusion dominates ($E_a \approx 0.8$–1.2 eV, Sect. 2.4.1). This explains why at present there is a lot of investment and interest in research for barrier materials, to boost the activation energy.

The interconnect surroundings affect not only the activation energy of the surface diffusion, but also the mechanical constraints. EM can be counteracted and stopped with stress migration, which is initiated by exposing the interconnect to mechanical stress. The dielectric has a greater role to play here than the barrier materials. It is only with sufficiently large mechanical tension gradients that an appreciable stress migration can take place. For this reason, dielectrics with high Young's moduli (i.e., high stiffnesses) produce the best results. The dielectric specified by the technology affects the EM response in this context.

In general, technological restrictions specify constraints for the layout design, which are typically referred to as *design rules*. These design rules result in specific local geometrical configurations that have implications for EM behavior. The overall design itself has very little impact on these configurations; instead, the technological specifications for the design are typically spacing, overlap, and width rules. There are often other rules as well, for example, for the surface ratios between metal and dielectric for every routing layer. Certain dimensions, such as the coating thickness of individual layers or the size of vias, are also typically specified by the technology.

2.3.2   Environment

We can see from Eq. (2.1) that the key environmental factor regarding electromigration is temperature and, as such, the physical location where the integrated circuit is deployed is critical. Some of the highest temperature standards apply for ICs in automobile electronics, where circuits are typically designed for ambient temperatures up to 175 °C (347 °F). These maximum temperatures can be reached in normal operations, in particular in gasoline engine compartments.

All electronic components and wire interconnects dissipate heat; this *power dissipation* is the difference between the energy supplied to an electrical component and that released during operation. High power losses, which frequently occur in digital circuits such as high-performance microprocessors or analog amplifiers, can cause increased temperature loading. This situation is further exacerbated if high power losses are combined with high ambient temperatures—this increases the likelihood of aging in integrated circuits.

Both of these temperature loads, from power losses and high ambient temperatures, reinforce EM by providing a portion of the activation energy as thermal energy. Furthermore, the diffusion process is accelerated by the increased mobility of ions. Copper is particularly susceptible to temperature changes; for example, if the operating temperature is increased by 10 K, the current needs to be cut by more than 50% to maintain the same median time to failure (service life duration). This critical relation can be derived with copper parameters from Black's Eq. (2.1). On the other hand, a 5 K decrease in operating temperature can lead to about a 25% increase in permissible current density (Fig. 2.7 illustrates this relationship for aluminum wiring).

The characteristic heat increase of the interconnect at high current densities due to Joule heating is another thermal factor that must be considered. As mentioned earlier, Joule heating is caused by interactions between the moving electrons and the metal ions that comprise the body of the conductor. Joule heating also leads to temperature gradients in the interconnect, which can cause thermal migration (Sect. 2.5).

Other environmental factors besides temperature also affect electromigration. For example, different substances can penetrate the metal layer by diffusion and impact the electromigration process. In addition, electrolytic migration (Sect. 2.1) can occur if water comes in contact with the wiring. Furthermore, oxygen also affects the processes by oxidizing the metals. These phenomena introduce major changes to the entire migration process that can rapidly cause severe damage. Such

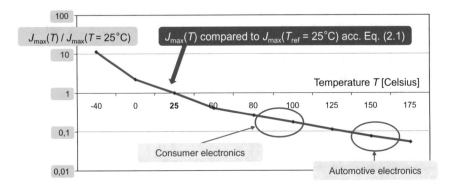

Fig. 2.7 Illustration of the relationship between maximum current density and temperature if the reliability *MTF* of an Al wiring in Eq. (2.1) is kept constant [Lie05, Lie06]. It becomes clear, that, for example, when the working temperature of an Al interconnect is raised from 25 °C (77 °F) to 125 °C (257 °F), the maximum tolerable current density must be reduced by about 90% in order to maintain the same reliability of the wire

environmental effects are not dealt with below as this type of diffusion of foreign substances can be prevented with suitable barriers.

2.3.3 Design

The design itself can significantly affect EM by defining the current densities that occur throughout the chip. The current density j, which is represented as the quotient of current and cross-sectional area (Eq. (1.1), Chap. 1), is determined at a specific location by the functional load, i.e., the electrical current I, and by the physical design solution, in particular, the cross-sectional area A of the interconnect at that location. The interconnect must be designed to deliver the currents required by the circuit; thus, the width of each interconnect must be adapted to accommodate the current.

Changes in direction and moving between layers cause local increases in current densities, which in turn compounds EM and leads to an agglomeration of damage (Fig. 2.8). All geometrical structures of the layout design, that is, those not exclusively specified by the technology, can be used to increase the EM-limited service life duration of the wiring, as illustrated in Fig. 2.8 by the use of suitable corner-bend angles.

Other factors that can reduce the impact of EM should be deployed as well, to prevent high local current densities. Service life duration can particularly benefit from limiting the length of the interconnects if the Blech length [Ble76] is leveraged. This is due to mechanical stress migration in interconnects that are less than a critical length, which counteracts EM and prevents damage occurring (Sects. 2.5 and 4.3).

Current density

Min.   Max.

Fig. 2.8 Current-density visualization of different corner-bend angles of a wire on an analog integrated circuit, *left* 90°, *middle* 135°, and *right* 150°. It shows that 90° corner bends must be avoided, since the current density in such a bend is significantly higher than that in oblique angles of, for example, 135°

In addition to current density, the frequency is another quantity arising from the design that impacts interconnect reliability. The change in direction of the current causes a corresponding change in the direction of diffusion, as well, and small (preexisting) damage to the interconnect can be partially cleared. This beneficial process is known as *self-healing* and greatly depends on the frequency of the current (Sects. 2.4.3 and 4.7).

2.4 Electromigration Mechanisms

As stated earlier, the dominant driving force for electromigration damage is due to momentum transfer from the moving electrons to the ions, which leads to a mass flux in the direction of the electron flow. The detailed mechanisms of this flux with regard to circuit materials, frequencies, and mechanical stresses are described next.

2.4.1 Crystal Structures and Diffusion Mechanisms

Interconnects in integrated circuits can have different crystal lattice structures (Fig. 2.9). The most common type of lattice structures in metallic interconnects is polycrystalline, fine-grained structures. Depending on the ratio between grain size and interconnect dimensions, there may also exist—in theory at any rate—poly-crystalline interconnects composed of few grains, bamboo-like structures, monocrystals, and amorphous structures. These categories for crystal lattice

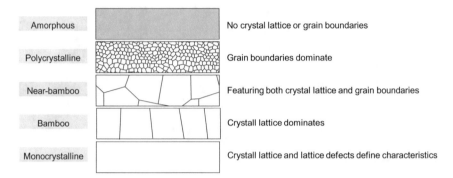

Amorphous	No crystal lattice or grain boundaries
Polycrystalline	Grain boundaries dominate
Near-bamboo	Featuring both crystal lattice and grain boundaries
Bamboo	Crystall lattice dominates
Monocrystalline	Crystall lattice and lattice defects define characteristics

Fig. 2.9 Different crystal lattice structures in metallic interconnects

structures are particularly useful for characterizing the causes and effects of electromigration. We discuss below the properties of these crystal lattice categories.

In amorphous interconnects, atoms are not in ordered structures, but rather are arranged in irregular patterns and hence have a short-range order, rather than a long-range one[3]. There are thus neither grain boundaries nor crystalline zones, and the material has very specific properties. Because there is no long-range order, there are no diffusion channels as in a periodic crystal lattice. In an amorphous lattice, the cohesion and atomic density differ from the crystalline state. In practical terms, we note that metals can only be brought into the amorphous state in extreme conditions, with cooling rates on the order of 10^5–10^6 K/s [SW96]. Such conditions cannot be created for the fabrication of integrated circuits, and thus the use of metal-based amorphous lattices is more of theoretical interest for EM mitigation.

The other extreme, the monocrystalline state, can also only be achieved with massive investment in technology, which too makes it impractical. Typically, the crystal has to be grown from a single germ in the molten mass. This is virtually impossible in an interconnect surrounded by different substances.

All the same, we should not completely rule out the use of monocrystalline lattice structures in the future. A variant of monocrystal growth, which works below the molten temperature, was used in [JT97]. Here, aluminum monocrystals were produced on a monocrystalline sodium chloride layer. Different crystal orientations can also be promoted with this process. This technology, however, is currently only suitable for use in the laboratory and not for IC fabrication, not least because of the unwanted sodium chloride in the semiconductor processes. The other issue with monocrystals is that individual lattice defects can greatly impact interconnect properties. This undermines the required process stability, which is so important for reliability.

---

[3]Long-range order in a crystal means that atoms are organized in a periodic order across many atoms, such as in a periodic lattice.

For these reasons, the polycrystalline state in interconnects is the norm. There are many different types of polycrystalline lattice structures whose properties differ enormously from an EM perspective. We can consider the bamboo and near-bamboo lattice structures (see Fig. 2.9) as polycrystalline lattice variants, whose diffusion properties are dominated by specific features, as we discuss below. Lattice diffusion (also called volume or bulk diffusion) as well as grain-boundary diffusion can take place in fine-grained polycrystalline microstructures (see Fig. 2.9). Surface diffusion occurs, too, regardless of the lattice; we will deal with this type of diffusion in Sect. 2.4.2.

EM, like all other types of migration, obeys the laws of diffusion. A crystal metal lattice can be modeled in terms of EM with the simplified one-dimensional diffusion formula (also known as heat equation[4]) for homogeneous media, as follows:

$$\frac{\partial c}{\partial t} = D \cdot \frac{\partial^2 c}{\partial x^2},$$ (2.2)

with the concentration c, the time t, the diffusion coefficient D, and the location x. The diffusion velocity v of the atoms, excited by the current density, can be expressed according to [AN91] as:

$$v = \frac{D}{kT} \cdot ez^* \varrho j.$$ (2.3)

In this equation, k is Boltzmann constant, T the absolute temperature, e the elementary charge, z^* the effective charge of the metal ion as a measure of the momentum exchange, ϱ the specific electrical resistance, and j the current density [AN91, Ble76].

The diffusion coefficient D expresses the magnitude of the atomic flux. It is a physical constant dependent on atomic size and other properties of the diffusing substance as well as on temperature and pressure. The diffusion coefficient is calculated in the case of the combined grain-boundary and bulk diffusion as follows:

$$D = D_v + \delta \cdot \frac{D_b}{d},$$ (2.4)

where D_v is the diffusion coefficient for the bulk (volume) diffusion and D_b the diffusion coefficient for the grain-boundary diffusion. The width of the grain boundaries δ and the mean grain size d must be considered, as well [AN91].

As shown in Table 2.1, different diffusion paths are characterized by different activation energies E_a (see Black's Eq. (2.1)). While the maximum activation energy is needed for bulk diffusion, it is lower for grain-boundary diffusion and surface diffusion. Accordingly, the diffusion coefficient D for bulk diffusion is

---

[4]The heat equation is a differential equation that describes the distribution of heat (or variation in temperature) in a given region over time.

Table 2.1 Activation
energies for different diffusion
paths for electromigration in
aluminum and copper

Diffusion process	Activation energy in eV	
	Aluminum	Copper
Bulk diffusion	1.2	2.3
Grain-boundary diffusion	0.7	1.2
Surface diffusion	0.8	0.8

smaller than for the other diffusion types. Hence, EM is more prevalent at grain boundaries and boundary layers. The material determines which of the two diffusion paths has the lowest activation energy.

If one ignores boundary effects and focuses on the core of an interconnect, one will see that it is primarily the grain boundaries that serve as diffusion paths. Hence, the density and direction of grain boundaries in the interconnect lattice significantly affect susceptibility to EM, and thus also the resulting reliability of the interconnects.

The link between grain size and electromigration damage was first detected at the end of the 1960s, and aggregate failures occurring at the transition between different grain sizes were measured [AR70]. Two-thirds of the defects found in aluminum strips were found to have occurred at the transition between extremely different grain sizes.

Polycrystalline lattice structures with a low grain-boundary density are potentially more robust to EM. Near-bamboo or bamboo-type structures (see Fig. 2.9) have fewer grain boundaries aligned in the direction of current flow. Grain-boundary diffusion can thus be partially stopped by using such variants.

Near-bamboo structures have individual crystallites—known as *blocking grains*—that expand across the entire width of the interconnect and inhibit the diffusion flux. However, damage tends to occur in the proximity of these crystallites as a result of void formation or material accumulation. This damage is caused by a divergence in the diffusion that occurs at these blockages or triple points. *Triple points* are points at which grain boundaries branch off, so that one grain boundary from one direction proceeds as two in other directions, or vice versa (Fig. 2.10).

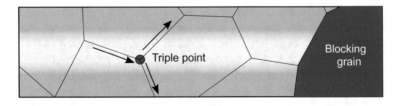

Fig. 2.10 Triple point(s) and blocking grain in a near-bamboo grain structure. In triple points, one grain boundary is split into two (or vice versa); blocking grains expand across the entire interconnect cross-section

2.4.2 Barriers of Copper Metallization

The use of copper interconnect has become dominant in recent years, but brings with it specific electromigration issues. Migration in copper wires is greatly affected by boundary effects due to the low activation energy for surface diffusion in copper (see Table 2.1).

Copper metallizations are primarily produced with *Damascene* technology (Fig. 2.11). This is a metal patterning process that can also be described as *additive patterning*. First, recesses, such as trenches (b) or via holes, are created in the previously deposited dielectric (a) in a lithographic process. Copper is then deposited on the wafer (c), so that the recesses are also filled. The wafer is then polished (d) by chemical-mechanical planarization (CMP), and the excess copper above the top edge of the recesses is removed. The interconnects and vias remain in the recesses.

Dual-Damascene technology can reduce the number of CMP steps involved: here, copper is deposited on an interconnect layer and a via layer underneath it in a single step. Hence, a trench and the underlying via may both be filled with a single copper deposition.

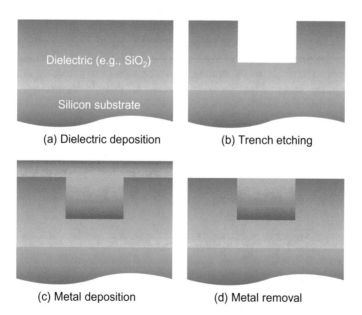

(a) Dielectric deposition   (b) Trench etching

(c) Metal deposition   (d) Metal removal

Fig. 2.11 A simplified schematic of the *Damascene* process on a cross-section of a copper track (**a**). The dielectric insulating layer is patterned with open trenches where the conductor is required (**b**). A coating of copper that significantly overfills the trenches is deposited on the insulator (**c**). Chemical-mechanical planarization (CMP) is used to remove the copper that extends above the top of the insulating layer (**d**). Copper sunken within the trenches is not removed and becomes the patterned conductor

It must be noted, however, that copper tends to diffuse considerably into neighboring silicon and silicon oxide at high temperatures [UON+96]. As a result, the above process is only beneficial if copper is treated with further technological measures.

In addition, temperatures on the order of 500 °C can be reached in the manufacture of the metallization—especially during the annealing process for creating bamboo structures (Sect. 4.2) [CS11]. The resulting diffusion has two major drawbacks: (i) a copper silicide layer with low conductivity is produced, and (ii) copper can degrade and destroy the semiconducting properties of silicon.

In order for circuits that incorporate copper metallization layers to function properly, *barriers* between metal and dielectric are essential. These diffusion barriers for copper and silicon must meet different criteria depending on their use. Good adhesion to copper and the dielectric, as well as thermal and mechanical stability with thin deposited layers (a few nanometers), are common criteria for these barriers.

The term *barrier* encompasses both the metallic diffusion barrier, the so-called *metal liner,* in the trench and the mostly dielectric protective coating, the *dielectric cap* (Fig. 2.12). There is therefore always a barrier between metal and dielectric (metal liner and dielectric cap) and between interconnect and the via above it (metal liner). This configuration is required to block diffusion especially during chip fabrication [UON+96].

The metal liner is deposited in the etched trench or via hole in the dielectric before the copper is deposited. The interconnect benefits from good electronic conductivity in the barrier, as the barrier layer (i.e., metal liner) is placed between the via and the underlying metal layer (see Fig. 2.12, right). In addition, the metal liner can also contribute to the current flow. This ensures a residual conductance especially in the event of faults arising from voids, thereby improving reliability.

The dielectric cap resides on top of the interconnects. Following the copper removal by CMP, a barrier to the above deposited dielectric is required. A dielectrical barrier is beneficial in preventing further structural modifications. A thin layer of dielectric cap material is deposited on the whole wafer prior to

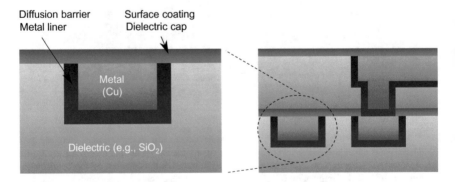

Fig. 2.12 Schematics of cross-section of copper tracks with the necessary, surrounding barrier layers (not to scale)

interlayer dielectric deposition. It needs only to be subsequently reopened when etching the via holes to make an electrical contact between the vias and the metal layer (see Fig. 2.12, right).

An electrically conductive cap would need to be structured lithographically, similarly to the underlying interconnect layer, in order to avoid shorts and parasitic conductance. Alternatively, the cap could be accumulated exclusively on the surface of the copper as a self-organizing process [CLJ08, LG09, VGH+12] and leaving out the exposed dielectric (Sect. 4.8.3).

With this better understanding of barrier construction and characteristics, we can now see, as noted earlier, that the barrier is a key factor for EM, as it forms a part of the boundary layer in copper metallizations that is critical for EM. Thus, the barrier greatly impacts the effective activation energy of the copper surfaces; we explore this below.

Theoretically, the activation energy of surface diffusion in the case of copper can be increased as well above the grain-boundary diffusion level by suitably selecting the barrier material, thus blocking surface diffusion. However, inhibiting one diffusion mechanism generally causes another mechanism to become predominant, leading to alternate damage scenarios. For example, the switch from aluminum to copper has eliminated grain-boundary diffusion, but it saw a significant increase in surface diffusion. Now, if surface diffusion is prevented by suitable dielectric and barrier layers, grain-boundary diffusion becomes an issue again. In the end, bulk diffusion may even emerge as the dominant process for electromigration, if all other mechanisms are suppressed. Every change in the dominant diffusion process therefore changes the failure modalities, as well, and increases the complexity of modeling procedures for EM prevention.

The difference between dielectric cap and metal liner also has a bearing on EM, as there are critical technological differences between the covered copper surfaces. In the dual-Damascene process [Gup09, Yoo08], a thick layer of copper is deposited on the wafer, and this layer is then removed by polishing (chemical-mechanical planarization, CMP). Copper is left only in the interconnect layer and underneath in the via layer. This process causes flaws in the surface of the metal, which also cannot be cleared in the process. The high defect density and vacancy concentration at the surface of the interconnects modify the surface characteristics at these locations and thus the activation energy. When combined with a worsening of adhesion of the dielectric cap, the top surface becomes more susceptible to electromigration damage, which is why voids typically occur on the top surface of an interconnect. These scenarios have implications on specific interconnect geometries (Sect. 4.4) and the materials used (Sect. 4.8).

2.4.3 Frequency Dependency of Electromigration

If the direction of the current in an interconnect is reversed, the direction of EM diffusion is also reversed. Due to this compensation by material backflow, damage caused by EM can be partially cleared. This effect is known as a *self-healing*, which can significantly extend the lifetime of a wire.

Whether damage can be effectively remediated by self-healing, thus contributing to the service life of an interconnect, depends on the amount of damage done and to what extent the crystal lattice has been changed before the current reverses direction. Frequency is therefore the key parameter at work here, as, along with the duty factor[5], it defines the duration of the one-sided current load.

Very little metal is moved per half cycle at high frequencies. Hence, there are very few changes to the microstructure. The current flow in the second half cycle is approximately a mirror image of the flow in the first half cycle, so that it is highly likely that the changes are reversed. This delays the first occurrence of damage in the form of vacancy defects and voids. Tests carried out at different frequencies show that an alternating resistance change (the self-healing component) is superimposed on a slowly rising resistance [TCH93]. Partial self-healing is thus verified.

As described in [TCH93, TCC+96], the scale of self-healing can be expressed with the diffusion fluxes J as follows:

$$J_{\text{net}} = J_{\text{forward}} - J_{\text{back}} = J_{\text{forward}} \cdot (1 - \gamma), \tag{2.6}$$

where γ is the self-healing coefficient. This coefficient is determined by the duty factor r of the current and other factors influencing the scale of self-healing, such as the frequency.

In [DFN06], the self-healing coefficient γ is introduced by expanding Black's Eq. (2.1) as follows:

$$\text{MTF}_{\text{AC}} = \frac{A}{\left(r \cdot j^+ - \gamma \cdot (1 - r) \cdot j^-\right)^n} \cdot \exp\left(\frac{E_a}{k \cdot T}\right). \tag{2.7}$$

The self-healing coefficient γ is determined empirically in the same publication by:

$$\gamma = \frac{r \cdot \frac{j^+}{j_{\text{DC}}} - s \cdot \frac{\text{MTF}_{\text{DC}}}{\text{MTF}_{\text{AC}}}}{(1 - r) \cdot \frac{j^-}{j_{\text{DC}}}}. \tag{2.8}$$

The current density of the positive half cycle is j^+, and j^- for the negative half cycle. The scaling factor s is determined iteratively.

Tao et al. [TCH93] found a median lifetime (median time to failure, MTF) increase over low frequencies (DC, for example) for copper interconnects by a factor of 500 for frequencies ranging from 10 to 10^4 Hz (Fig. 2.13). Rectangular wave current signals were used in this study at frequencies ranging from a few mHz

---

[5]A duty factor is the fraction of one period in which a signal or system is active, i.e., it expresses the ratio of the positive pulse duration to the period. The duty factor is commonly scaled to the maximum of one. A duty cycle expresses the same notion; however, it is labeled as a percentage.

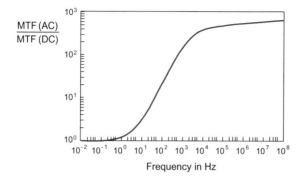

Fig. 2.13 Median time to failure (MTF) if interconnect is stressed by an alternating current (AC) compared to MTF if a directed current (DC) is applied [TCH93]. Note the almost linear increase in reliability when the frequency increases over several orders of magnitude (cf. Fig. 4.30, Chap. 4)

to 200 MHz. The reason for the limited lifetime of interconnects even at high frequencies, where we might hope for self-healing to extend the useful lifetime infinitely, is the interaction between EM and thermal migration, which degrades and destroys the interconnects. We discuss such interactions in Sect. 2.5.

It must be said, however, that the change in lifetime occurs in a frequency range that has very little relevance for today's digital circuits. Signals in this "low" frequency range are mainly handled by subcircuits that deal with the environment or the human-machine interface (Table 2.2).

Other signal frequencies are generally much higher (mega or gigahertz), while currents in supply lines consist of a dominant DC component with superimposed harmonics at very low frequencies. Hence, high frequencies on their own are not enough to prevent damage. The frequency dependency does, however, show that we need to differentiate between signal lines and power supply lines when dealing with EM.

An empirical model for healing damage is developed in [DFN06]; see Eq. (2.7). Shono et al. [SKSY90] also modeled the forward and backward transportation of metal due to the reversal in current flow. They assume that the amount of charge in both directions is the same (that is, there is no DC component), but that the current waveform is asymmetric w.r.t. time. While there are long current pulses of low

Table 2.2 Examples of relevant frequencies

Example	Frequency
Controlling the background lighting for a computer screen	10 mHz
Frame rate on a PC monitor	60 Hz
Sampling frequency for audio signals	44 kHz
Carrier frequency for radio frequency identification (RFID)	13.56 MHz
Processor clock frequency	3 GHz

amplitude in one direction, current pulses in the opposite direction are shorter with larger amplitudes. Hence, there is an asymmetrical material transport with a net flux in one direction due to the nonlinear relation between material transport and current density. The minimum lifetime is reached with the model at a duty factor, that is, the ratio of the positive pulse duration or pulse width (*PW*) to the period (*T*), of approximately 0.4.

Having shown the positive effects of alternating currents on reliability, we must also point out one drawback. Current is displaced from the current-carrying conductor at very high frequencies due to a phenomenon known as the *skin effect*[6]. At such high frequencies, the wire interior contributes very little to the current flow, which causes the current density to increase at the outer regions of the wire. A measure of the current displacement, the skin depth δ, is given by:

$$\delta = \sqrt{\frac{2\varrho}{\omega \cdot \mu}}, \tag{2.9}$$

where ϱ is the specific electrical resistance and μ the magnetic permeability of the conductor material. The variable ω represents the angular frequency with $\omega = 2\pi f$.

The current density decreases approximately exponentially, with the variable j_S representing equivalent surface current density and d as the distance from the surface, as follows:

$$j \approx j_S \cdot \exp\left(-\frac{d}{\delta}\right). \tag{2.10}$$

A better approximation of the current-density distribution as a function of the radius r in a long cylindrical conductor with current I can be analytically expressed as follows:

$$j_{\text{eff}}(r) = \frac{I}{2\pi r_0} \cdot \sqrt{\omega\kappa\mu} \cdot \sqrt{\frac{r_0}{r}} \cdot \exp[-\sqrt{\omega\kappa\mu} \cdot (r_0 - r)]. \tag{2.11}$$

In this formula from [KMR13], the electrical conductivity is represented by $\kappa = 1/\varrho$ and the cross-sectional radius of the conductor by r_0. As it stands, this analytical derivation of the current density cannot be applied to conductors with rectangular cross-sections. The mathematical model of a cylindrical conductor in Eq. (2.11) suffices here as an estimation of the magnitude.

In the case of copper, the skin depth at 50 Hz is approximately 9.4 mm and is proportional to $1/\sqrt{f}$. A critical frequency of 90 GHz was determined for the skin

---

[6]The skin effect is due to opposing eddy currents induced by the changing magnetic field resulting from the alternating current. This effect leads to a reduction in current from the outside to the inside of a metallic conductor as a function of the frequency and the electrical material constants of the conductor (permeability and conductivity).

Fig. 2.14 A comparison between minimum structure sizes and skin depth in relation to the skin effect; data from [ITR14]. The graph shows that the skin effect can be ignored in future, as well, in the lower metallization layers for a semiconductor scale of 100 nm and less

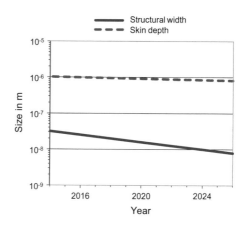

effect for an interconnect of square cross-section with dimensions width W and height t with $W = t = 0.45$ μm in [WY02]. Using a similar calculation, the critical frequency of approximately 35 THz for the 22-nm technology node was found to be much higher.

Problems arising from the skin effect are not expected in digital circuits in the light of current developments in semiconductors with regard to track widths and clock frequencies [ITR14, ITR16]. The reason for this is that the interconnect dimensions are being downscaled more quickly than the frequency-dependent skin depth (Fig. 2.14).

We note that at present the skin effect can disturb the high-frequency signal components and thus the clock edges. Furthermore, the skin effect is reduced at lower conductivities and is thus weakened by the increasing influence of boundary effects on the interconnects.

2.4.4 Mechanical Stress

There are three main causes of mechanical tension (mechanical stress) in interconnects:

- The metal is deposited at high temperatures (approximately 500 °C) [CS11]. Mechanical stress is induced by the cooling to ambient temperature due to the different thermal expansion coefficients of metal and insulator.
- The growth of layers during metal deposition is generally uneven, which also produces mechanical stress in the metallization. This issue is more critical than the first effect according to [CS11]; the phenomenon can be illustrated by wafer curvature measurements [CS11, BLK04].

- The material transport caused by EM redistributes vacancies in the crystal metal lattice which, in turn, generate mechanical stress. A vacancy represents an absence of atoms; the absence of an atom needs less room than an atom at the same lattice site. Hence, the lattice can relax, leading to a local reduction in material volume and lower compressive stress.

The nature of the mechanical stress in interconnects can differ depending on the combination of materials and the production process. Damage is typically caused by tensile or compressive stress in the interconnect, which lead to a failure mechanism. Smaller tensions can be relieved by lattice dislocations and typically do not produce failure mechanisms.

As noted above, mechanical stress results from the fabrication of interconnects due to different thermal expansion coefficients and high temperatures during metal deposition. The temperature of the unstressed state, around 250 °C [ZYB+04], is generally significantly higher than the maximum operating temperature. Interconnects are exposed to mechanical tensile forces at standard operating temperatures, as the thermal expansion coefficient of copper, at 16.5×10^{-6} K^{-1} [Gup09], is much larger than the surrounding dielectric (SiO$_2$: 0.5×10^{-6} K^{-1} [YW97]).

Using the parameters Young's modulus E (a measure of the stiffness of a solid material) and temperature T

- $E(Cu) = 117$ GPa,
- $E(SiO_2) = 70$ GPa,
- $\Delta T = 200$ K,

and assuming identical widths of metal and dielectric (Fig. 2.15) for a one-dimensional calculation with the approximation:

$$\frac{\sigma}{E} = \varepsilon = \alpha \cdot \Delta T, \qquad (2.12)$$

where ε is the strain and α the thermal expansion coefficient, we obtain a tensile stress σ of almost 140 MPa in the horizontal direction perpendicular to the longitudinal direction of the interconnect with:

$$\sigma = \frac{\alpha_{SiO_2} - \alpha_{Cu}}{\left(E_{SiO_2}^{-1} + E_{Cu}^{-1} \right)} \cdot \Delta T. \qquad (2.13)$$

Tensile stresses promote the creation of voids. The modified topology resulting from the formation of voids tends to relieve tensile stresses, and the region at the edge of the voids typically becomes stress-free. Although voids may seem "beneficial" for their ability to relieve mechanical stress, void formation should be strongly avoided, as the mechanical contact between metal and dielectric is destroyed and the conductive cross-section of the interconnect is reduced.

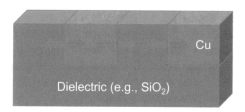

Fig. 2.15 Cooling of copper wires embedded in dielectric (silicon dioxide) leads to tensile stress, marked by arrows at the interfaces, due to differences in coefficients of thermal expansion (CTE)

The allowed compressive stresses in interconnects are usually greater than the allowed tensile stresses. If, however, a critical compressive stress threshold is exceeded, this also leads to a reduction in tension. In this case, interconnect extrusions are formed that spread into the neighboring dielectric as *dendrites, whiskers,* and *hillocks* (see Fig. 2.2). This is comparable to the transition from elastic to plastic deformation in solid mechanics.

Before extrusions arise, the vacancy concentration is scaled back further; this process is partially reversible through, e.g., self-healing (Sect. 2.4.3). Extrusions, however, are irreversible and lead to severe damage to the circuit, or its destruction.

2.5   Interaction of Electromigration With Thermal and Stress Migration

In addition to EM, there are three other types of diffusion in metallic connectivity architectures that can significantly impact reliability: thermal migration, stress migration, and chemical diffusion. IC designers must be especially aware of thermal and stress migration; both are introduced and described in their interaction with EM in this section.

Temperature gradients produce *thermal migration*. Here, high temperatures cause an increase in the average speeds of atomic movements. The number of atoms diffusing from areas of high temperature to areas of lower temperature is higher than the number diffusing in the opposite direction. As a result, there is a net diffusion in the direction of the negative temperature gradients, which can lead to significant mass transport.

Stress migration describes a type of diffusion that leads to a balancing of mechanical stress. Whereas there is a net atomic flow into areas where tensile forces are acting, metal atoms flow out of areas under compressive stress. Similar to thermal migration, this leads to diffusion in the direction of the negative mechanical tension gradient. As a result, the vacancy concentration is balanced to match the mechanical tension.

Chemical diffusion occurs in the presence of a concentration (or chemical potential) gradient, which also results in a net mass transport. This type of diffusion is always a nonequilibrium process; it increases system entropy[7] and brings the system closer to equilibrium. Since chemical diffusion is quite different from the migration processes mentioned above and does not directly relate to EM in metallic interconnects, it is not discussed further here.

Migration processes can lead to an equilibrium state, where the limiting (or counteracting) process is always another type of migration. There can be an equilibrium between electromigration and stress migration (the so-called *Blech effect*), between thermal migration and chemical diffusion (the so-called *Soret effect*), or any other combination of two or more migration types.

2.5.1 *Thermal Migration*

Temperature gradients produce thermal migration (TM), sometimes also referred to as *thermomigration*. Here, high temperatures cause an increase in the average speeds of atomic movements. Atoms in regions of higher temperature have a greater probability of dislocation than in colder regions due to their temperature-related activation. This causes a larger number of atoms diffusing from areas of higher temperature to areas of lower temperature than atoms in the opposite direction. The result is net diffusion (mass transport) in the direction of the negative temperature gradients (Fig. 2.16).

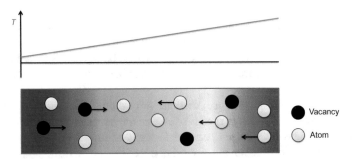

Fig. 2.16 Thermal migration (TM) is expressed by atomic and vacancy movement. It consists of mass transport from one local area to another, much like EM, with the difference that TM is driven by a thermal gradient rather than an electrical potential gradient (T temperature)

---

[7]Entropy is a measure of the "disorder" of a system. Hence, the more "ordered" or "organized" a system is, the lower its entropy. For example, building blocks that have been used to construct a wall are "highly organized" (i.e., they are arranged in a complex structure) and are thus in a *low-entropy* state. This state is achieved only by the input of energy. If this structure is left unattended, it will decay after a number of years, and the disorganized, high-entropy state will return (i.e., an unorganized heap of blocks).

The main reasons for temperature gradients in metal wires are

- Joule heating inside the wire caused by high currents,
- external heating of the wire, such as caused by highly performant transistors nearby,
- external cooling of the wire, which may result from through silicon vias (TSV) connected to a heat sink, in connection with low thermal conduction of the wire and its surrounding, such as through narrow wires surrounded by a thermally insulating dielectric.

Interestingly, thermal migration also contributes to thermal transport, as heat is coupled to the transported atoms. This means that thermal migration directly moderates its own driving force, which contrasts with EM, where current density is only indirectly reduced by increased resistance in some cases.

If the temperature gradients are beyond the control of the thermal migration, i.e., the equilibrium state of minimal energy and homogeneous temperature cannot be reached, a steady state can still be attained. In this case, migration is stopped by linear gradients of other migration processes' driving forces and entropy is generated by the heat flow [Soh09].

Thermal migration is very prominent in metal alloys such as solders, where migration dissolves the alloy due to different mobilities of the alloy components (the Soret effect) [Soh09]. Here, thermal migration and chemical diffusion set up an equilibrium. This, and the fact that temperature gradients are higher in packaging applications, makes thermal migration an interesting field of research especially in solder connections, such as flip-chip contacts.

The process has less influence in interconnect structures located within integrated circuits, as almost pure metals and no alloys are used, and the temperature gradients are tempered by the high thermal conductivities of metal and insulation.

Thermal migration has been an important field of study in solder joints, for it is likely to happen during regular operation. A temperature differential of 10 K across a flip-chip contact of 100 μm diameter creates a temperature gradient of 1000 K/cm, which suffices to induce thermal migration in the solder [Tu07].

2.5.2 Stress Migration

Stress migration (SM), sometimes also referred to as *stress voiding* or *stress-induced voiding* (SIV), describes atomic diffusion that leads to a balancing of mechanical stress. There is a net atomic flow into areas where tensile forces are acting, whereas metal atoms flow out of areas under compressive stress. Similar to thermal migration, this leads to diffusion in the direction of the negative mechanical tension gradient (Fig. 2.17). As a result, the vacancy concentration is balanced to match the mechanical tension.

The main reasons for mechanical stress as the driving force behind SM in metal wires are thermal expansion, electromigration, and deformation through packaging.

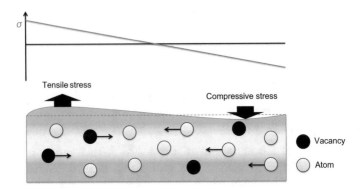

Fig. 2.17 Stress migration is a result of a mechanical stress gradient, either from external forces or from internal processes, such as electromigration or thermal expansion. Voids form as a consequence of vacancy migration driven by the hydrostatic stress gradient (σ mechanical stress)

A mismatch of the thermal expansion coefficients between metal, dielectric, and die material, and the temperature change from fabrication to storage, as well as the working conditions, cause most of the stress. By using TSVs for contacting 3D-stacked ICs, this initial stress is increased and it becomes less uniform, as well.

Metal lattices usually contain vacancies, i.e., some of the atomic positions in the lattice are unoccupied. Although they are aligned with the lattice grid, vacancies consume less space than atoms at the same positions. Therefore, the volume of a crystal that contains vacancies is to some extent smaller than the volume of the same crystal with atoms in the place of former vacancies. Vacancies play a major role in stress migration. Via Hooke's Law (which states that the strain or deformation of an elastic object is proportional to the stress applied to it):

$$\sigma = E \cdot \varepsilon \tag{2.14}$$

this volume is coupled with mechanical stress. Here, σ and ε are the mechanical stress and the strain, respectively, while E is Young's modulus. The change in volume (strain in three dimensions) correlates to an inverse pressure change. If the number of vacancies is reduced, pressure or compressive stress increases. The decline or increase in the number of vacancies is caused by the place change of atoms.

The stress gradient drives atoms from high pressure regions to regions with tensile stress and pushes vacancies in the opposite direction. This effect is equivalent to a highly viscous fluid that reacts slowly to an external pressure gradient. The external stress gradient is minimized in this case by structural deformation. Initially, microscopic atomic or vacancy motion facilitate this process. Temperature has a critical effect on the process, as it enables the "place changing" of atoms, which, in turn, causes vacancies to move.

In the case of external mechanical stress, the crystal lattice is stretched or compressed depending on the kind of stress. While there is an increased likelihood of atoms migrating to the stretched regions, atoms in the compressed regions are "pushed" outward to increase the number of vacancies; the required volume and the stress are thus reduced (Fig. 2.18). The result is an atomic flux from regions of compressive stress to regions of tensile stress until a static state with no stress gradient is reached.

If the stress is exerted *internally* by migration processes, e.g., by EM, there will be a greater concentration of vacancies in regions of tensile stress. This concentration will be balanced by stress migration to a steady state, where the atomic flux due to EM is compensated by SM.

If the number of vacancies induced by *external* stress or EM exceeds a threshold, the vacancies unite to form a void due to vacancy supersaturation. This phenomenon is often called *void nucleation*. Subsequently, the tensile stress is reduced to zero by the resulting crack [HPL+10]. At the same time, the driving force for SM changes, as well as the equilibrium state (Fig. 2.19).

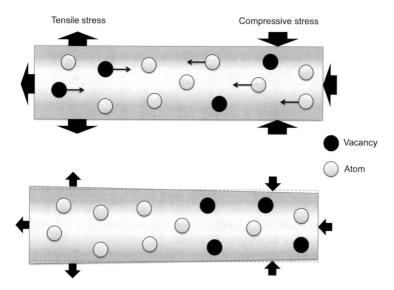

Fig. 2.18 Stress migration leads to diffusion of atoms and vacancies (*top*) to eliminate the origin of this migration (*bottom*). Atoms migrate into the stretched regions (left-hand side, outward-facing stress arrows), whereas atoms in the compressed regions are "pushed" outward (right-hand side, inward-facing arrows). Note that this material flow from compressive to tensile stress is in the opposite direction compared to the EM flow

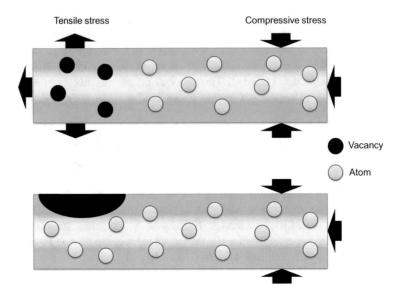

Fig. 2.19 Vacancy supersaturation (*top*) leads to the formation of voids (*bottom*), also called void nucleation. Note that the resulting crack at *bottom* eliminates the (external) tensile stress

2.5.3   Mutual Interaction of Electromigration, Thermal and Stress Migration

Electromigration (EM) interacts directly with stress migration (SM), as the dislocation of metal atoms induces mechanical stress, which is the driving force behind SM. SM works against EM, as its flow is directed from compressive to tensile stress which is the opposite direction to the EM flow. The resultant net flow is thus reduced, and the damaging dislocation due to EM is slowed or even halted.

Thermal migration (TM) on the other hand is not a dedicated EM countermeasure, as it is less dependent on the current direction than EM. Its direction can differ from the EM direction depending on the temperature gradient, which might stem from sources other than current density.

While temperature accelerates EM as well as the other migration types, we observe most likely a mixture of all three types in the event of a current-density hotspot. For the effective application of countermeasures, the dominant migration force must be identified.

EM, TM, and SM are closely coupled processes as their driving forces are linked with each other and with the resultant migration change (Fig. 2.20).

Current density increases the temperature through Joule heating, and temperature change modifies mechanical stress through differences in the expansion coefficients. Furthermore, temperature and mechanical stress affect the diffusion coefficient [see Eq. (2.20)], which in turn modifies the behavior of all three migration types. In

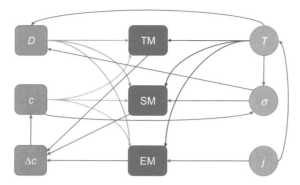

Fig. 2.20 Interaction and coupling between electromigration (EM), thermal migration (TM), and stress migration (SM) through their driving forces current density (j), temperature (T), and mechanical stress (σ). Also shown are the migration parameters diffusion coefficient (D), concentration (c), and concentration change (Δc), respectively

addition, the mechanical stress is influenced by the change in atomic concentration caused by all migration types individually.

The effects of different combinations of the three main migration types are depicted in Figs. 2.21 and 2.22. Depending on the origins of the driving forces, several different amplifying and compensating results are observed.

The causes and effects of migration are interrelated and at times self-reinforcing. For example, recall our earlier discussion on void growth, current density, and Joule

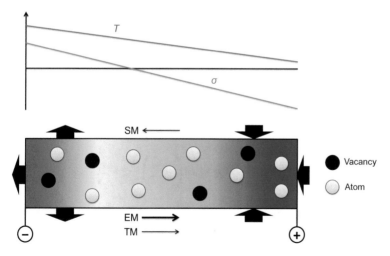

Fig. 2.21 Example of coupled migration processes in a wire segment, where electromigration and thermal migration proceed from left to right, while the stress migration flow moves in the opposite direction (T Temperature, σ mechanical stress)

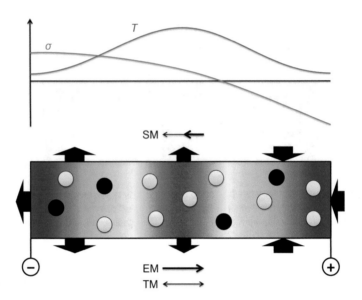

Fig. 2.22 Another example of coupled migration processes. Here, thermal migration is induced through a hotspot in the middle of the segment, while the stress is a superposition of tensile stress in the middle and EM-induced stress. This situation may occur near thermal vias or TSVs

heating in reference to Fig. 2.4, where we illustrated the acceleration of void growth by the positive feedback of a temperature rise. In general, the effects of different migration modes should be considered as interdependent. In particular, the material flows J_E from EM, J_T from thermal migration, and J_S from stress migration can be calculated as follows [WDY03]:

$$\overrightarrow{J_E} = \frac{c}{kT} \cdot D_0 \cdot \exp\left(-\frac{E_a}{kT}\right) \cdot z^* e \varrho \, \overrightarrow{j}, \qquad (2.15)$$

$$\overrightarrow{J_T} = -\frac{cQ}{kT^2} \cdot D_0 \cdot \exp\left(-\frac{E_a}{kT}\right) \cdot \text{grad } T, \qquad (2.16)$$

$$\overrightarrow{J_S} = -\frac{c\Omega}{kT} \cdot D_0 \cdot \exp\left(-\frac{E_a}{kT}\right) \cdot \text{grad } \sigma. \qquad (2.17)$$

In these equations, c is the concentration of atoms, k Boltzmann's constant, T the absolute temperature, D_0 the diffusion coefficient at room temperature, E_a the activation energy, z^* the effective charge of the metal ions, e the elementary charge, ϱ the specific electrical resistance, j the electrical current density, Q the transported heat, Ω the atomic volume, and σ the mechanical tension (stress).

The resultant diffusion flux, defined as follows:

$$\overrightarrow{J_a} = \overrightarrow{J_E} + \overrightarrow{J_T} + \overrightarrow{J_S},\qquad(2.18)$$

is the net effect of the combined driving forces.

The individual diffusions can flow in the same or in opposite directions. There is also a coupling of the effects, that is, the feedback effect of the diffusion on the causes of the material transfer, which should not be ignored. For example, the critical length effects, covered later in this book (Sect. 4.3), arise from this type of negative feedback between EM and SM.

The resulting diffusion flow in one dimension is described in [Tho08] as follows:

$$J_a = \frac{Dc}{kT} \cdot \varrho j z^* e + \frac{Dc}{kT} \cdot \Omega \cdot \frac{\partial \sigma}{\partial x},\qquad(2.19)$$

where J_a is the atomic flux, D the diffusion coefficient of copper, represented by:

$$D = D_0 \cdot \exp\left(-\frac{E_a}{kT}\right),\qquad(2.20)$$

c is the concentration of copper atoms, j the current density, z^* the effective charge of copper, e the elementary charge, k the Boltzmann's constant, T the temperature, Ω the atomic volume of copper, σ the mechanical tension (stress), and x the coordinate parallel with the segment with $x = 0$ at the cathode.

In order to prevent EM effects, the net diffusion flow must be reduced to zero. This means that the diffusion flow from EM and, for example, the corresponding diffusion flow from SM (in the opposite direction) may be used to cancel each other out.

2.5.4 Differentiation of Electromigration, Thermal and Stress Migration

The particular damage arising from a given migration type cannot be identified by appearance, as all damage, regardless of its root cause, results in voids caused by diffusion processes (Fig. 2.23). However, the locations and surroundings of these different damage types provide evidence as to their possible origin(s) (Fig. 2.24).

Diffusion barriers are key in all diffusion processes considered here because damage will most likely occur near such barriers, due to flux divergences and the effects of bad cohesion.

As discussed earlier, EM takes place inside wires and is driven by electric currents. Therefore, EM damage correlates mostly with the current direction and strength. EM damage is most likely to be found in areas of high current density, that is, high currents and small cross-sectional areas. In addition, current crowding at wire bends and vias is a strong EM indicator.

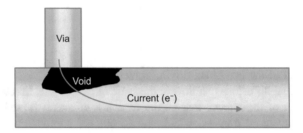

Fig. 2.23 Visualization of damage caused by the combined effect of EM, SM, and TM (side view)

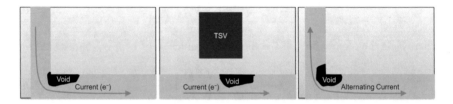

Fig. 2.24 Different types of damage typically caused by EM (*left*), SM (*middle*), and TM (*right*) (top view). In most cases, the respective damage cannot be differentiated by its appearance, but rather by its location and surroundings

TM correlates somewhat with EM, as large temperature differences appear near locations of high current densities. Therefore, current crowding spots are also high temperature spots that are a potential TM driver. Here, large temperature differences (in addition to current differences) "push" the atoms.

There are many other reasons for temperature gradients, such as external heating or cooling and the heating of active circuit elements, namely transistors. Furthermore, thermal conduction influences temperature gradients. Thermal conduction can also dislocate TM damage from hotspots in the wires toward cooling spots or areas of low thermal conductivity. This might be a TM indicator, whereas EM is always coupled to large current locations.

Another EM characteristic is its requirement of a directed current flow. Wires with alternating current flow, such as digital signal lines, often show TM as a partial source of damage growth (in addition to EM, see Fig. 2.24, right).

Due to the prevailing combination of different materials in an interconnect system, the resulting temperature inhomogeneities always lead to mechanical stress. Therefore, TM is mostly coupled with SM, with SM often being the dominant force. In order to apply appropriate countermeasures, we need to know whether large temperature gradients can occur inside the region of interest, or if the migration is driven by stress gradients only. These different migration scenarios require different countermeasures.

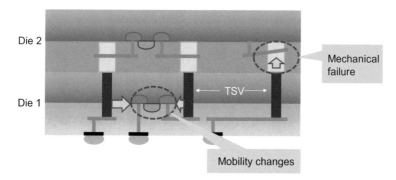

Fig. 2.25 TSVs in 3D-stacked ICs induce mechanical stresses in their surroundings because of the mismatch in the coefficients of thermal expansion between silicon and copper, and other effects [XK11]. This results in mechanical failures and electrical degradation, such as mobility changes in transistors. As one precaution, keep-out zones around TSVs are recommended

SM is often coupled with EM in terms of counteraction. EM-transported atoms induce mechanical stress that consequently leads to SM in the opposite direction to the causal EM. SM therefore has the potential to reduce EM damage in short wire segments (Sect. 4.3) and in locations of low current densities.

SM due to mechanical stress not only originates from EM, but also from fabrication, mismatches between different coefficients of thermal expansion (CTEs), and induced stress from obstacles like TSVs. With the increase in 3D-IC applications, damage near structures such as TSVs in 3D-stacked ICs is rapidly becoming critical. In most cases, it is SM related: TSVs induce stress on their surroundings due to the mismatch in the coefficient of thermal expansion values between silicon ($\alpha_{Si} \sim 3 \times 10^{-6}$ K^{-1}) and copper ($\alpha_{Cu} \sim 16.5 \times 10^{-6}$ K^{-1}) as a TSV fill. The resulting mechanical stress leads to, among others things, silicon wafer cracking, debonding between wafers and TSV protrusion, signal degradation and cracking (Fig. 2.25) [XK11]. Hence, to successfully implement TSVs, the mechanical stresses in the copper TSV itself as well as in the surrounding silicon substrate must be controlled. As one precaution, keep-out zones around TSVs are created for active devices to minimize their stress-related mobility changes[8] [KML12].

Finally, we would like to point out that hillocks and whiskers (see Fig. 2.2) usually indicate EM as their cause. However, SM can also participate in the overall diffusion flow and, hence, must be considered as well.

---

[8]Electron mobility is a measure of how quickly an electron can move through a material such as a metal or semiconductor, when pulled by an electric field.

2.6   Migration Analysis Through Simulation

This last section of this chapter introduces and describes the principles of a migration analysis through simulation. We discuss EM analysis using current-density simulation techniques, such as the finite element method (FEM), in Sect. 2.6.1. We also outline more sophisticated simulation strategies in the following subsections. For example, the atomic flux can be calculated from current density and other driving forces to get a deeper insight into the migration process (Sect. 2.6.2). We may also simulate mechanical stress development as the driving force behind stress migration (Sect. 2.6.3). Void growth can be simulated in order to gain a detailed look into damaging processes (Sect. 2.6.4).

2.6.1   Simulation Techniques

Migration is a complex problem that can be described by a system of differential equations. For this type of mathematical problem, several solving strategies exist and can be classified as follows:

- analytical methods,
- quasi-continuous methods,
- concentrated or lumped element methods, and
- meshed geometry methods, such as

    - finite element method (FEM),
    - finite volume method (FVM), and
    - finite differences method (FDM).

All these methods must respect numerous boundary conditions given by the simulation problem as well as the coupling effects of the different physics domains participating in the migration process. The solution space consists of a set of variables, also called "degrees of freedom," which must be adjusted to fit the boundary conditions and equations.

The system of differential equations can only be solved *analytically* in closed form and with acceptable effort for very simple geometries and boundary conditions, or by extremely simplifying the problem by neglecting some transport processes and simplifying the geometry, for example.

To facilitate the solution space, *quasi-continuous methods*, such as "power blurring" [ZPA+14], have been developed. In contrast to concentrated elements, this method provides a spatially resolved solution by superposition of analytical expressions for a finite number of spatial points, like a grid. It uses a matrix convolution technique similar to image processing methods to combine the separate point solutions to a global solution. This approach consumes less computational power than meshed solutions while losing some of the flexibility, as it is harder to implement inhomogeneous material properties. For temperature calculations, as

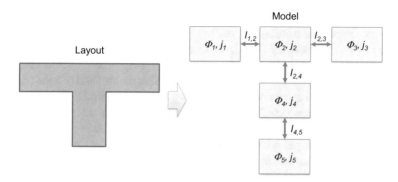

Fig. 2.26 Illustration of the lumped element model where several analytically or numerically solved geometries are combined (Φ electrostatic potential, j current density, I current)

used in power blurring, quite reasonable results can be obtained, even for full-chip analysis [KYL15].

A *concentrated or lumped element method* (Fig. 2.26) is really an extension of the analytical method, where several analytically or numerically solved geometries are combined. This method is very fast, but calculates only a single value for each degree of freedom and element. The results are global without any spatial resolution inside the segments, as the elements are typically quite large, e.g., one element per wire segment.

Meshed geometry methods (Fig. 2.27) offer several advantages for migration analysis. The degrees of freedom can be spatially resolved in a variable manner by adjusting the mesh granularity. The calculation effort is limited due to the bounded degrees of freedom—the mesh is finite. Using only basic geometries for the mesh elements further simplifies the simulation.

The *finite element method* (FEM) is a universal tool for calculating elliptic and parabolic equation systems. It is a numerically very robust method. Many tools

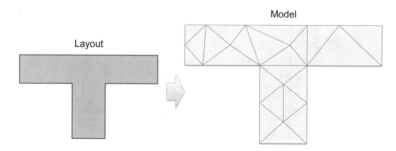

Fig. 2.27 Illustration of the meshed model where the spatial resolution of the degrees of freedom can be adjusted by the mesh granularity. Hence, each mesh node has its own degrees of freedom, such as the electrostatic potential

support FEM due to its great variety of applications. The system of equations is built from degrees of freedom for nodes and elements.

The *finite volume method* (FVM) uses polyhedrons to divide the given geometry, while solving the equations only at the center of each polyhedron. FVM is best suited for conservational equations, such as mass flow calculations for fluid and gas transport. It could be applied to migration when modeling atomic flux similar to gas diffusion.

The *finite differences method* (FDM) is numerically very simple and therefore well suited for theoretical analysis or very fast calculations. Due to its simplicity, its results are not as exact as with FEM. As its name suggests, the system of equations is based on the differences in the degrees of freedom.

All these methods are regularly deployed for solutions in computational fluid dynamics (CFD), which has a lot in common with migration simulation in solid state.

For reduced problem sizes, such as EM analysis restricted to power and ground nets, meshed methods deliver precise results in reasonable calculation times. However, when applying meshed methods to complex geometries, model preparation and calculation efforts are extremely high. These issues apply also in EM analysis, as geometries in VLSI circuits are becoming increasingly complex. Since signal nets are more and more EM-affected, filtering only EM-critical nets, as proposed in [JL10], will no longer sufficiently curb problem complexity. This growing challenge of complexity for finite element simulations for EM problems in VLSI circuits is clearly shown in Fig. 2.28.

Quick simulations are required in physical design. These simulations are only one part of the verification phase; they must be repeated iteratively in the design flow. For example, applying FEM for use in the full-chip verification of complex integrated circuits is too time consuming [TBL17].

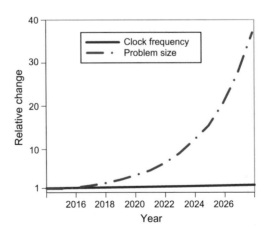

Fig. 2.28 Prediction of the analysis complexity for future digital circuits, i.e., shown is the growing problem size of finite element simulations of all signal nets in future technologies, as predicted by the ITRS, with 2014 as the base year. The respective CPU clock frequencies are also depicted for comparison purposes. Calculated from ITRS [ITR14, ITR16]

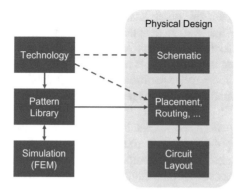

Fig. 2.29 To mitigate increasing circuit complexity, an EM simulation based on FEM should be uncoupled from the actual verification process. The resulting layout synthesis would then use a pattern verification method that restricts physical design to pre-verified (routing) patterns [TBL17]

In order to maintain FEM precision despite the increasing number of structures and geometries to be analyzed, we propose that EM simulations must be separated from the actual verification process. This means that FE analysis is performed prior to verification or even prior to routing. Routing will then be implemented exclusively with verified routing elements from a library. Hence, a whole library of routing elements with simplified parametric models attached will be verified in an FE analysis. The complete chip can then be verified rapidly. The library should include all routing elements required to build a complete layout; the library size can be minimized by selecting highly repetitive patterns. The verification is simplified to calculating only actual critical results from the actual boundary conditions by using the parametric models to check against current-density limits, or other migration metrics (Fig. 2.29).

An important prerequisite for the above-mentioned verification method of combining several discrete FEM simulations is that the partial solutions equate with the respective parts of the complete solution. This requirement is fulfilled if the boundary conditions are transformed correctly between the full and partition models, as we discuss next.

The method's prerequisite can be best explained in a typical example, where a complete wire connection is simulated as a whole and then split into separate parts. If we can transform the boundary conditions to the parts in a suitable manner, the simulation results will be equivalent.

There are several useful rules for finding the locations to split the model. The best place to split is at locations where the boundary conditions are homogeneous, as they can easily be applied to FE models. This is the case, for example, with current-density regions in a straight wire that are located at some distance from vias and branches. If, however, the layout element of interest consists only of the via region, some appendices will have to be added to the wires in order to establish a homogeneous boundary condition.

The atomic flux, on the other hand, stops at diffusion barriers, that is, the transition from one material to another (this typically occurs near vias). These diffusion barriers offer ideal boundary conditions for this model.

In the case of mechanical stress, it may not be easy to find ideal locations for dividing the model into smaller partitions because mechanical stress is also driven by the wire's surroundings; a homogeneous stress condition is difficult to find around a wire.

Temperature influences and mechanical stress from "unmodeled" surroundings should not influence the simulation results. Therefore, a larger wire surrounding area should be modeled, so that the difference between homogeneous model conditions and inhomogeneous real conditions can be neglected inside the wire.

The same applies when modeling temperature directly, as the surrounding dielectric distributes heat as well as the metal, only with lower conductivity.

In summary, it is important to verify that FE routing models can be partitioned without losing accuracy if one wants to apply FEM for full-chip current-density analysis. This verification is best done by comparing the simulation results for generic sample patterns calculated both separately and combined. Figures 2.30, 2.31, and 2.32 visualize this using a T-shape wire segment inside one metal layer and a via connection. Figure 2.30 shows the current-density results from two separate (distinct) simulations. The simulation of both patterns combined is pictured in Fig. 2.31; the combined results correspond well with the separately calculated results. Figure 2.32 indicates current-density distribution at the interface between the two patterns in a joint simulation; this is a measure of the error in the individual simulations. The maximum error is 3% in the visualized case; this is an acceptable value that has been verified for other patterns as well [TBL17].

By pre-evaluating interconnect structures in advance and building the layout exclusively from evaluated structures, verification is significantly accelerated: even a single circuit simulation, i.e., generating the (simulated) library patterns and using

Fig. 2.30 Current-density distribution using FEM for a T-shape wire segment and a via connection. Shown are results of separate simulations of the two single patterns with homogeneous constraints at the cut surfaces

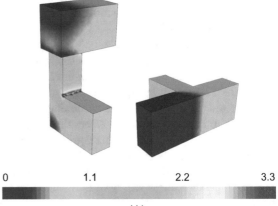

| 0 | 1.1 | 2.2 | 3.3 |

j / j_0

Fig. 2.31 Visualizing the joint FEM current-density simulation of the two patterns in Fig. 2.30 when combined indicates sufficient similarity with the results of the disjoint simulations (cf. Fig. 2.30)

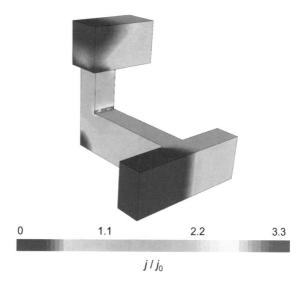

$j \, / \, j_0$

Fig. 2.32 Verifying homogeneity of the current density at the cut surface between the two FEM submodels (the maximum deviation is 3% here) can be used to ensure that distinct and combined simulations show matching results. Hence, FE wiring models can be partitioned without accuracy loss if one wants to apply FEM for full-chip current-density analysis

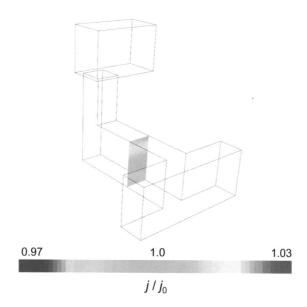

$j \, / \, j_0$

them only once, can be faster than a conventional, complete FEM simulation of the entire final layout [TBL17].

The aforementioned method of pre-verifying routing patterns allows FEM, including its precision and spatial resolution, to be applied in the EM verification of complex circuits. All circuit nets can be verified by following the proposed method. We believe this is a particularly important result, as it will fill a looming gap in the

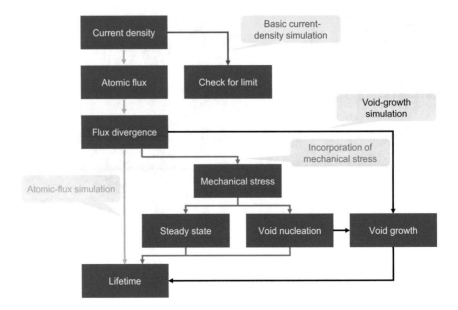

Fig. 2.33 Simulation strategies for EM analysis based on different parameters affecting migration

near future when all segments will be severely EM-affected (Chap. 1, cf. Fig. 1.6, red area).

So far we have discussed EM analysis using current-density simulation and a subsequent comparison with current-density limit(s). There are also several other, more sophisticated simulation strategies for EM analysis (Fig. 2.33); these are outlined in the following subsections. For example, the atomic flux can be calculated from current density and other driving forces to get a deeper insight into the damaging process (Sect. 2.6.2). Furthermore, we can simulate mechanical stress development as the driving force behind stress migration and compare it with the critical stress (Sect. 2.6.3). Void growth can be simulated in order to gain a detailed look into damaging processes, in terms of both void nucleation (mechanical stress change) and void growth (Sect. 2.6.4).

2.6.2 Atomic-Flux Simulation

The diffusing atom flux is used to quantify the rate of diffusion. The flux is defined as either the number of atoms diffusing through a unit area per unit time (atoms/ $(m^2 \cdot s)$) or the mass of atoms diffusing through a unit area per unit time (kg/$(m^2 \cdot s)$).

The atomic flux can be calculated by solving the systems of equations for all migration driving forces and deriving the sum of all fluxes. In our case of EM analysis, the migration driving forces are the current density for electromigration, the temperature gradient for thermal migration, and the mechanical stress gradient for stress migration.

In this section, we explain how to calculate the atomic flux when the driving force is known. Numerous models are available for solving this task. In the following discussion, we go from the smallest scale to more abstract models.

The most natural approach is to calculate the *movement of single atoms*, also called atomistic simulation. In the case of interconnect structures, this yields a statistical model or, in small scale, a stochastic material transport model. The driving force or work W_p is translated into a material flux or mean velocity $\bar{v}$ by relating it with an energy barrier and the atomic mobility. The resistance of an atom to motion is given by its binding energy E_b and its mass, or mobility m. This leads to the exemplary equation as follows:

$$\bar{v} = \left(W_p - E_b\right) \cdot \frac{1}{m}. \tag{2.21}$$

Single atomic migration can be calculated on this basis. Probabilistic calculations must be made in order to obtain an atomic flux from this approach.

A more abstract method of calculating atomic flux uses collective atomic properties to calculate a *mean flux*. Here, statistics over a certain number of atoms are included in the model equation.

The most abstract model uses Eqs. (2.15)–(2.20) from Sect. 2.5.3 to calculate an atomic flux from the driving forces of the migration types. This deterministic model gives only mean flux results.

The atomic flux can be calculated in a quasi-static simulation (Fig. 2.34) which determines the initial atomic flux and the spatial flux divergence. Hence, critical regions can be directly identified in the locations of large divergences. Lifetime and robustness can then be estimated by extrapolating this flux.

The microstructure has a significant influence on EM, as diffusivity differs for bulk and grain boundaries [COS11]. Hence, different local EM properties are

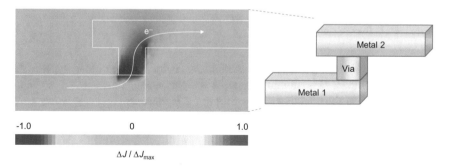

Fig. 2.34 Atomic flux divergence calculated from electromigration (FEM results) shows via depletion (side view)

changed due to microstructural variations. Furthermore, mechanical properties depend on crystal orientation—an important factor in SM as an EM countermeasure. Therefore, the microstructure must be included in the geometric models.

The microstructure can be incorporated in the geometric models by using a microstructure generator to establish a structure from median grain size and a given standard deviation, as noted in [COS11]. As microstructure and crystal orientation cannot be completely controlled by process technology, this method gives results for one arbitrary configuration only. In reality, however, the positions of grain boundaries and the crystallographic orientations cannot be arranged. Hence, deterministic methods will not yield realistic simulation results. A probabilistic analysis of the microstructure effects is, therefore, required for reliable outcomes [COS11].

The divergences can be deduced from the atomic-flux simulation results; these allow the lifetime of the layout structure, for example of a wire, to be estimated by extrapolation.

2.6.3 Simulation of Mechanical Stress

In addition to simulating atomic flux due to EM, it is mandatory to calculate the mechanical stress caused by atomic flux, as it is key to determining the steady-state condition in the metal wire. In particular, the total flux in short wires greatly depends on mechanical stress.

The obtained steady state leads either to an "immortal" wire (cf. Sect. 4.3) or the void size in this state gives an indication as to the wire's lifetime.

As metal wires in integrated circuits are confined, i.e., encapsulated, by dielectric material, every material movement annihilates and/or generates vacancies. There are different models of mechanical stress generation and evolution due to this process, such as the Korhonen model[9] [Ye03]. It describes the stress by the one-dimensional equation as follows:

$$\frac{\partial \sigma}{\partial t} = \frac{\partial}{\partial x}\left[\frac{D_a B \Omega}{kT}\left(\frac{\partial \sigma}{\partial x} - \frac{z^* e \varrho}{\Omega}j\right)\right], \tag{2.22}$$

where σ is the hydrostatic stress, t is the time, D_a is the atomic diffusivity, B is the applicable modulus, Ω is atomic volume, k is Boltzmann's constant, T is absolute temperature, z^* is the effective charge number, e is electron charge, ϱ is the resistivity, and j is the current density [Ye03].

---

[9]The Korhonen model combines vacancy dynamics with stress development. It assumes that the recombination and generation of vacancies alter the concentration of the available lattice sites, which influences the hydrostatic stress distribution. Specifically, the loss of the available lattice sites increases the hydrostatic stress.

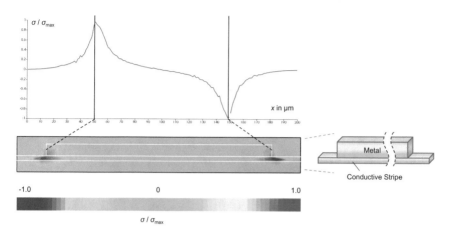

Fig. 2.35 Simulation results for mechanical stress through EM after a fixed stressing time. The nonlinear stress gradient inside the wire is shown (side view)

In addition, several similar and more sophisticated models exist for three-dimensional calculations [Ye03].

Having determined the mechanical stress due to EM, overstepping the critical stress threshold indicates where void nucleation starts [CS11]. In the case of a transient EM simulation to this point in time, the nucleation time and, depending on the dominant failure mechanism, the wire lifetime can be estimated.

This failure criterion—void nucleation—is equivalent to a small resistance increase in experiments, according to [CS11]. Simulation results can thus be experimentally verified.

EM lifetimes vary widely due to high stress thresholds and large variations in grain distribution [CS11]. Hence, probabilistic calculations are necessary here as well.

To illustrate these mechanical stress considerations, simulation results are shown in Figs. 2.35 and 2.36. Figure 2.35 illustrates the mechanical stress buildup after a fixed simulation time, and Fig. 2.36 depicts a steady-state condition in a short wire.

2.6.4   Void-Growth Simulation

Void growth can be simulated to provide deeper insight into the electromigration processes and the effects that lead to failure. We can thus estimate the lifetime of different layout structures in a very detailed fashion. In general, void-growth simulation is the third step in a migration analysis, following the estimation of driving forces and the atomic flux calculation. Therefore, it incorporates relatively complex mathematics to calculate its results.

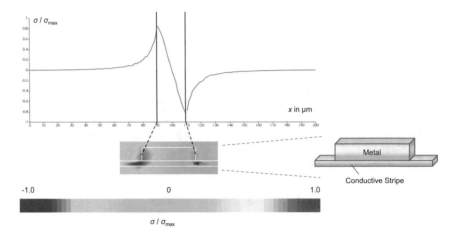

Fig. 2.36 Simulation results for a short-length segment including mechanical stress. The steady state with linear stress profile inside the segment is shown (side view)

After having calculated the (static) atomic flux, actual atomic *motion* increases the vacancy concentration in certain locations and voids are created by vacancy supersaturation. This process and the change of the void shape must be modeled as well in order to estimate void growth and, hence, the lifetime of a layout structure. A transient simulation is necessary to calculate the void growth.

As there are many parameters in the complex calculations and some assumptions may be necessary, the results may not be reliable. We also need to conduct a statistical analysis on input parameter variations to make useful conclusions based on the simulation results.

When considering void nucleation it is important to note that during EM, and in combination with mechanical stress, a significant grain-boundary movement takes place [CS11]. Here, grain boundaries drift into the neighboring crystallite lattice (grain) as a result of atomic rearrangements. Under these circumstances, one grain grows at the expense of another. This process must be considered as well, as it influences the overall diffusion flow and causes electrical resistance fluctuations.

The applicability of finite element models for simulating migration processes and void growth until failures occur has been shown in [BS07] and [THL07].

Two methods are available for modeling void growth in a meshed geometry:

(1) geometry modification depending on volumetric loss of affected elements (Fig. 2.37), similar to a method from [OO01], and
(2) deletion of mesh elements upon exceeding a certain mass flux divergence limit (Fig. 2.38), as presented in [WDY03].

Both methods must also include surface tension models to generate the energy-based void shape modification. The aims of these models are to gain a deeper understanding of void growth and, thus, to identify methods for lifetime

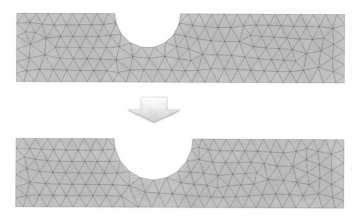

Fig. 2.37 Void-growth model using mesh geometry modification [OO01]

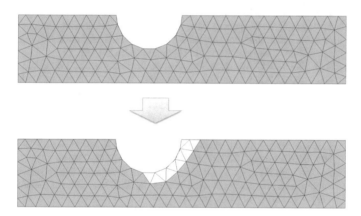

Fig. 2.38 Void-growth model using deletion of mesh elements [WDY03]

extension. The latter can be achieved by modifying wire geometries or by implementing special reservoirs; both methods are presented in detail in Chap. 4.

References

[AN91]   E. Arzt, W.D. Nix, A model for the effect of line width and mechanical strength on electromigration failure of interconnects with 'near-bamboo' grain structures. J. Mater. Res. **6**(4), 731–736 (1991). https://doi.org/10.1557/JMR.1991.0731

[AR70]   M.J. Attardo, R. Rosenberg, Electromigration damage in aluminum film conductors. J. Appl. Phys. **41**(6), 2381–2386 (1970). https://doi.org/10.1063/1.1659233

[Bl69a]   J.R. Black, Electromigration—a brief survey and some recent results. IEEE Trans. Electron Devices **16**(4), 338–347 (1969). https://doi.org/10.1109/T-ED.1969.16754

[Bl69b]   J.R. Black, Electromigration failure modes in aluminum metallization for semicon-
           ductor devices. Proc. of the IEEE **57**(9), 1587–1594 (1969). https://doi.org/10.1109/
           PROC.1969.7340
[Ble76]   I.A. Blech, Electromigration in thin aluminum films on titanium nitride. J. Appl.
           Phys. **47**(4), 1203–1208 (1976). https://doi.org/10.1063/1.322842
[BLK04]   W.G. Breiland, S.R. Lee, D.D. Koleske, Effect of diffraction and film-thickness
           gradients on wafer-curvature measurements of thin-film stress. J. Appl. Phys. **95**(7),
           3453–3465 (2004). https://doi.org/10.1063/1.1650882
[BS07]   A.F. Bower, S. Shankar, A finite element model of electromigration induced void
           nucleation, growth and evolution in interconnects. Modell. Simul. Mat Sci. Eng. **15**
           (8), 923–940 (2007)
[CLJ08]   H. Chang, Y.-C. Lu, S.-M. Jang, Self-aligned dielectric cap. U.S. Patent App. 11/
           747,105, 2008
[COS11]   H. Ceric, R. de Orio, S. Selberherr, Integration of atomistic and continuum-level
           electromigration models, in *18th IEEE International Symposium on the Physical
           and Failure Analysis of Integrated Circuits (IPFA)* (2011), pp. 1–4
[CS11]   H. Ceric, S. Selberherr, Electromigration in submicron interconnect features of
           integrated circuits. Mater. Sci. Eng.: R: Rep. **71**(5–6), 53–86 (2011). https://doi.org/
           10.1016/j.mser.2010.09.001
[DFN06]   L. Doyen, X. Federspiel, D. Ney, Improved bipolar electromigration model, in *44th
           Annual., IEEE International Reliability Physics Symposium Proceeding* (2006),
           pp. 683–684. https://doi.org/10.1109/relphy.2006.251323
[FWB+09]   R.G. Filippi, P.C. Wang, A. Brendler, et al., The effect of a threshold failure time
           and bimodal behavior on the electromigration lifetime of copper interconnects, in
           2009 IEEE International Reliability Physics Symposium (2009), pp. 444–451.
           https://doi.org/10.1109/irps.2009.5173295
[Gup09]   T. Gupta, *Copper Interconnect Technology*. Springer (2009). https://doi.org/10.
           1007/978-1-4419-0076-0
[Hau04]   C.S. Hau-Riege, An introduction to Cu electromigration, Microel. Reliab. **44**, 195–
           205. https://doi.org/10.1016/j.microrel.2003.10.020 (5, 2004)
[HPL+10]   A. Heryanto, K.L. Pey, Y. Lim, et al., Study of stress migration and electromigration
           interaction in copper/low-k interconnects, in *IEEE International Reliability Physics
           Symposium (IRPS)* (2010), pp. 586–590
[ITR14]   International Technology Roadmap for Semiconductors (ITRS), 2013 edn. (2014),
           http://www.itrs2.net/itrs-reports.html. Last retrieved on 1 Jan 2018
[ITR16]   International Technology Roadmap for Semiconductors 2.0 (ITRS 2.0), 2015 edn
           (2016), http://www.itrs2.net/itrs-reports.html. Last retrieved on 1 Jan 2018
[JL10]   G. Jerke, J. Lienig, Early-stage determination of current-density criticality in
           interconnects, in *Proceeding of International Symposium on Quality in Electronic
           Design (ISQED)* (2010), pp. 667–774. https://doi.org/10.1109/isqed.2010.5450505
[JJ11]   P. Jain, A. Jain, Accurate estimation of signal currents for reliability analysis
           considering advanced waveform-shape effects, in *24th International Conference on
           VLSI Design (VLSI Design)* (2011), pp. 118–123. https://doi.org/10.1109/vlsid.
           2011.61
[JT97]   Y.-C. Joo, C.V. Thompson, Electromigration-induced transgranular failure mech-
           anisms in single-crystal aluminum interconnects. J. Appl. Phys. **81**(9), 6062–6072
           (1997). https://doi.org/10.1063/1.364454
[KMR13]   K. Küpfmüller, W. Mathis, A. Reibiger, *Theoretische Elektrotechnik/Eine
           Einführung*, 19. aktual. Aufl. (Springer Vieweg, 2013). ISBN 978-3-642-37940-6
[KML12]   J. Knechtel, I.L. Markov, J. Lienig, Assembling 2-D blocks into 3-D chips, IEEE
           Transaction on Computer-Aided Design of Integrated Circuits and Systems, vol. 31,
           no. 2 (2012), pp. 228–241. https://doi.org/10.1109/tcad.2011.2174640

[KYL15]   J. Knechtel, E.F.Y. Young, J. Lienig, Planning massive interconnects in 3D chips, IEEE Transaction on Computer-Aided Design of Integrated Circuits and Systems, vol. **34**, no. 11 (2015), pp. 1808–1821. https://doi.org/10.1109/tcad.2015.2432141

[LG09]   A.R. Lavoie, F. Gstrein, Self-aligned cap and barrier. U.S. Patent App. 12/165,016, 2009

[Lie05]   J. Lienig, Interconnect and current density stress—an introduction to electromigration-aware design, in *Proceeding of 2005 Interconnect Workshop on System Level Interconnect Prediction (SLIP)* (2005), pp. 81–88. https://doi.org/10. 1145/1053355.1053374

[Lie06]   J. Lienig, Introduction to electromigration-aware physical design, in *Proceeding of International Symposium on Physical Design (ISPD)* (ACM, 2006), pp. 39–46. https://doi.org/10.1145/1123008.1123017

[LT11]   W. Li, C.M. Tan, Black's equation for today's ULSI interconnect electromigration reliability—A revisit, in *International Conference of Electron Devices and Solid-State Circuits (EDSSC)* (2011), pp. 1–2. https://doi.org/10.1109/edssc.2011. 6117717

[OO01]   T.O. Ogurtani, E.E. Oren, Computer simulation of void growth dynamics under the action of electromigration and capillary forces in narrow thin interconnects. J. Appl. Phys. **90**(3), 1564–1572 (2001). https://doi.org/10.1063/1.1382835

[SKSY90]   K. Shono, T. Kuroki, H. Sekiya, et al. Mechanism of AC electromigration, in *Proceeding Seventh International IEEE VLSI Multilevel Interconnection Conference* (1990), pp. 99–105. https://doi.org/10.1109/vmic.1990.127851

[Soh09]   Y. Sohn, Phase-field modeling and experimentation of constituents redistribution in metallic alloys, Slides of NIST Diffusion Workshop (2009), https://www.nist.gov/ sites/default/files/documents/mml/msed/thermodynamics_kinetics/NIST-09- Workshop-fsrd.pdf. Last retrieved on 1 Jan 2018

[SW96]   W. Schatt, H. Worch (ed.), *Werkstoffwissenschaft*, 8. neu bearb. Aufl. (Dt. Verl. für Grundstoffindustrie, Stuttgart, 1996). ISBN 3-342-00675-7

[TCC+96]   J. Tao, J.F. Chen, N.W. Cheung, et al., Modeling and characterization of electromigration failures under bidirectional current stress. IEEE Trans. Electron Devices **43**(5), 800–808 (1996). https://doi.org/10.1109/16.491258

[TCH93]   J. Tao, N.W. Cheung, C. Hu, Metal electromigration damage healing under bidirectional current stress. IEEE Electron Device Lett. **14**(12), 554–556 (1993). https://doi.org/10.1109/55.260787

[THL07]   C.M. Tan, Y. Hou, W. Li, Revisit to the finite element modeling of electromigration for narrow interconnects, J. Appl. Phys. **102**(3), 033705-1–033705-7 (2007)

[TBL17]   M. Thiele, S. Bigalke, J. Lienig, Exploring the use of the finite element method for electromigration analysis in future physical design, in *Proceeding of the 25th IFIP/ IEEE International Conference on Very Large Scale Integration (VLSI-SoC)* (2017), pp. 1–6. https://doi.org/10.1109/VLSI-SoC.2017.8203466

[Tho08]   C.V. Thompson, Using line-length effects to optimize circuit-level reliability, in *15th International Symposium on the Physical and Failure Analysis of Integrated Circuits (IPFA)* (2008), pp. 1–48. https://doi.org/10.1109/ipfa.2008.4588155

[Tu07]   K.-N. Tu, *Solder Joint Technology–Materials, Properties, and Reliability*. Springer (2007). https://doi.org/10.1007/978-0-387-38892-2

[UON+96]   M. Uekubo, T. Oku, K. Nii, et al., Wn_x diffusion barriers between Si and Cu. Thin Solid Films **286**(1–2), 170–175 (1996). https://doi.org/10.1016/S0040-6090(96) 08553-7

[VGH+12]   S. Van Nguyen, A. Grill, T.J. Haigh, Jr., et al., Self-aligned composite M-MOx/ dielectric cap for Cu interconnect structures. U.S. Patent 8,299,365, 2012

[WY02]   W. Wu, J.S. Yuan, Skin effect of on-chip copper interconnects on electromigration. Solid-State Electron. **46**(12), 2269–2272 (2002). https://doi.org/10.1016/S0038-1101(02)00232-0

[WDY03]   K. Weide-Zaage, D. Dalleau, X. Yu, Static and dynamic analysis of failure locations and void formation in interconnects due to various migration mechanisms. Mater. Sci. Semicond. Process. **6**(1–3), 85–92 (2003). https://doi.org/10.1016/S1369-8001 (03)00075-1

[XK11]   X. Xu, A. Karmarkar, 3D TCAD modeling for stress management in through silicon via (TSV) stacks. AIP Conf. Proc. **1378**(53), 53–66 (2011). https://doi.org/10.1063/ 1.3615695

[Ye03]   H. Ye, C. Basaran, D.C. Hopkins, Numerical simulation of stress evolution during electromigration in IC interconnect lines. IEEE Trans. Compon. Packag. Technol. **26**(3), 673–681 (2003). https://doi.org/10.1109/TCAPT.2003.817877

[Yoo08]   C.S. Yoo, *Semiconductor Manufacturing Technology*, World Sci. (2008). ISBN 978-981-256-823-6

[YW97]   X. Yu, K. Weide, A study of the thermal-electrical- and mechanical influence on degradation in an aluminum-pad structure. Microelectron. Reliab. **37**(10–11), 1545–1548 (1997). https://doi.org/10.1016/S0026-2714(97)00105-4

[ZPA+14]   A. Ziabari, J.-H. Park, E.K. Ardestani, et al., Power blurring: fast static and transient thermal analysis method for packaged integrated circuits and power devices, in *IEEE Transactions on VLSI Systems*, vol. 22, no. 11 (2014), pp. 2366–2379. https:// doi.org/10.1109/tvlsi.2013.2293422

[ZYB+04]   C.J. Zhai, H.W. Yao, P.R. Besser, et al., Stress modeling of Cu/low-k BEoL— Application to stress migration, in *Proceeding 42nd Annual IEEE International Reliability Physics Symposium* (2004), pp. 234–239. https://doi.org/10.1109/relphy. 2004.1315329

Chapter 3
Integrated Circuit Design and Electromigration

In this chapter, we will introduce measures for modifying the present integrated circuit (IC) design methodology with the objective of countering electromigration. After introducing the overall design flow (Sect. 3.1) in use today, we will explore how this flow is differentiated between analog and digital design, as both areas require different measures to counter electromigration (Sect. 3.2).

Understanding that knowledge of the currents flowing in interconnects is a fundamental requisite for an electromigration-aware design, we will discuss in Sect. 3.3 the different types of currents encountered and show how sensible current values can be determined.

As we gain further understanding of electromigration in integrated circuits, we will see that the key parameter for electromigration prevention is the maximum permissible boundary value of the current density in the wires. This parameter is, however, dependent on the intended use of the IC, which is why so-called mission profiles are created to determine such values. Section 3.4 describes how robust current-density boundary values can be determined, using application and reliability specifications.

Effective current-density verification is at the core of electromigration-aware design flows. This is highlighted in Sect. 3.5 where we delve deeper into the fundamental procedures and critical aspects of it. In Sect. 3.6, we present options for dealing with problems identified by current-density verification, using layout adjustment techniques.

In the final Sect. 3.7, we put forward a number of farther-reaching measures for increasing current-density boundary values, based on our assessment of current technological trends. These are then explored in further detail in Chap. 4.

© Springer International Publishing AG 2018
J. Lienig and M. Thiele, *Fundamentals of Electromigration-Aware
Integrated Circuit Design*, https://doi.org/10.1007/978-3-319-73558-0_3

3.1 Design Flow of Integrated Circuits

Our goal is to develop an IC design flow that is electromigration aware; as such, it is important to first understand the overall IC design, identify areas in the flow where inserting "electromigration awareness" is most beneficial, and determine where information to support such electromigration awareness comes from in the design flow (i.e., data dependencies among flow components). Thus before considering an electromigration-aware design flow, let us outline the main steps involved in designing an IC.

This highly complex process is divided into distinct tasks that are performed sequentially (Fig. 3.1). While early steps in the design flow are high level and operate on large pieces of the design in terms of logical inputs and outputs, later steps in the flow are performed at much lower levels of abstraction, typically involving detailed physical and electrical information. At the end of the process, before fabrication, design tools and designers use detailed information on each circuit element's geometric shape and electrical properties [KLMH11].

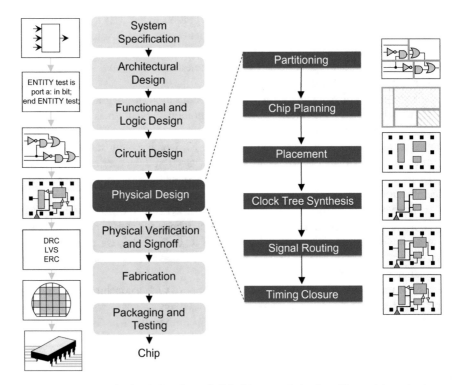

Fig. 3.1 Main steps in the design flow of digital integrated circuits with special emphasis on layout synthesis (physical design) [KLMH11]

Chip architects and designers, product marketers, and layout and library designers, collectively define the overarching goals and high-level requirements for the IC during *system specification*. These goals and requirements span functionality, performance, physical dimensions, and production technology.

The basic circuit architecture is then drafted to meet the system specifications during *architectural design*. Among the considerations here are packaging type and die-package interface, power requirements, and the choice of process technology.

Having decided on the architecture, the functionality and connectivity of each module are determined next. Only the high-level behavior is described during *functional and logic design*. Specifically, each module has a set of inputs, outputs, and timing behavior that the designer defines in this step.

The logic synthesis tool automatically converts Boolean expressions into a gate-level netlist for most of the digital logic on the chip, usually at the granularity of standard cells or higher. However, a number of critical, low-level elements ought to be designed at the transistor level; this is referred to as *circuit design*. The result of circuit design, as well as of logic synthesis, is a netlist, i.e., a description of the connectivity of the electronic circuit.

The main step covered in this book is *physical design*, where all design components are instantiated with their geometric representations. That is, all macros, cells, gates, transistors, etc., with fixed shapes and sizes per fabrication layer are assigned spatial locations on the IC (placement) and have appropriate interconnects embedded in the IC's metal layers (routing).

As physical design is a highly complex process, it is typically decomposed into several key steps (see Fig. 3.1, right) [KLMH11]:

- *Partitioning* subdivides a large circuit into smaller subcircuits or modules, which can each be designed and analyzed individually.
- *Chip planning* generally comprises two activities:

  - *Floorplanning* determines the geometries and arrangement of subcircuits or modules, as well as the locations of external ports and IP or macro blocks; and
  - *Power and ground routing*, often integral to floorplanning, distributes power (VDD) and ground (GND) nets throughout the chip.

- *Placement* establishes the spatial locations of all cells within each module (block).
- *Clock network synthesis* specifies the buffering, gating, and routing of the clock signal to meet prescribed skew and delay requirements.
- *Signal routing* is generally comprised of two sequential activities:

  - *Global routing* allocates routing resources to connections; and
  - *Detailed routing* assigns routes to specific metal layers and routing tracks within the global routing resources.

- *Timing closure* optimizes circuit performance by specialized placement and routing techniques.

When the physical design is completed, the layout must be fully verified to ensure correct electrical and logical functionality. Physical verification usually includes the following verification methods:

- *Design rule checking* (DRC) verifies that the layout meets all technology-imposed constraints.
- *Layout versus schematic* (LVS) derives a netlist directly from the layout and compares it with the original netlist produced from logic synthesis or circuit design.
- *Parasitic extraction* verifies the electrical characteristics of the circuit.
- *Antenna rule checking* seeks to prevent so-called antenna effects during manufacturing.
- *Electrical rule checking* (ERC) verifies the correctness of power and ground connections, and that signal transition times and other electrical characteristics are appropriately bounded.

The final, verified layout, usually represented in the GDSII or OASIS stream format, is sent for *fabrication* at a dedicated silicon foundry. Here, the design is patterned onto different layers using photolithographic processes. ICs are manufactured on round silicon wafers. At the end of the manufacturing process, the ICs are separated, or diced (into *die*), by precisely sawing the wafer into smaller pieces.

Functional chips are typically *packaged* and retested when dicing is complete. Packaging is configured early in the design process; it is application-oriented and embodies cost- and form-factor considerations. A die is positioned in the package cavity, and its pins are connected to the package's pins. Finally, the package is sealed and shipped to the customer as a working chip.

3.2 Electromigration-Aware Design Flows

An *electromigration-aware design* flow is recommended to avoid electromigration issues that may cause the chip to fail during operation. Here, measures to prevent electromigration (EM) are taken during the design steps outlined in Sect. 3.1. The measures can be implemented in individual design steps, or as measures that cut across multiple steps. EM can be taken into account during the physical design step, for example, by current-driven wire routing and adapting interconnect widths. It is imperative that the layout is also verified for EM robustness.

When drafting the layout during physical design, placement and/or signal routing should be performed to suit the expected currents, with the aim of reducing EM. The currents acquired from the circuit simulation and the permissible current densities are typically used as constraints in this context. The approach here is to place the function blocks at locations so as to minimize current flows, and to size the interconnect widths (routing widths) to suit these currents.

Interconnect widths and the wiring route (routing) are critical, especially for nets with more than two nodes, that is, for multi-terminal interconnects that span multiple pin or port connections. Currents can often be reduced in individual segments of this multi-terminal interconnect by selectively setting *Steiner points* [LJ03, LKL+12]. Steiner points are additional nodes in the wiring net, without any direct pin or port connection.

The current in the segments or the local current density is validated during verification, depending on the respective design step. The current density in the layout can be determined as a function of the geometrical characteristics of the routing pattern, and then compared with the maximum permissible current density of the specific routing layer. Any overshoots must be flagged and remedied with layout modifications.

The differences between analog and digital design, and their respective design methodologies, require different measures to effectively counter electromigration. There are three types of integrated circuits: (i) digital, (ii) pure analog, and (iii) mixed-signal; the latter contains digital and analog subcircuits. Designing analog and digital circuits are two very different activities.

Analog circuits are generally less complex and designed manually. Many functional constraints must be considered in analog design, which has held back design automation and analog synthesis to date and necessitates manual design. In addition, the design steps for digital circuits are typically discrete and are performed sequentially; analog design steps tend to overlap and several steps are often performed simultaneously. For example, device generation, module placement, and routing are normally executed simultaneously in analog designs.

Digital circuits in contrast are notable for their large number of nets and their fully automated design flows. While the complexity of analog circuits generally does not exceed a few thousand nets, digital circuits typically have millions of nets.

Further, digital circuits are characterized by net classes that exhibit different susceptibilities to electromigration [Ma89]. We can assume that power nets carry fairly constant currents and that their current directions are consistent. Clock and signal nets, on the other hand, conduct alternating currents (AC). As outlined in Sect. 2.4.3 (Chap. 2), these alternating currents impact wire lifetimes less than direct currents, due to self-healing mechanisms. Hence, the different net classes encountered in digital circuits require different current-density limits.

3.2.1   Analog Design

Analog circuits are designed as single modules, with a verified layout produced for each such module. The design process itself is only minimally automated; an engineer typically drafts the layout for an analog circuit (and analog parts in mixed-signal designs) almost completely manually.

The low degree of automation as compared to digital design is a result of the large number of degrees of freedom, influencing factors, and constraints. Circuit topology, component parameters, and drawing up the circuit layout are some of the degrees of freedom in the design. The performance and robustness of an analog circuit are negatively impacted by many factors, including non-linearities, parasitic components, electromagnetic coupling, temperature, and electromigration.

Constraints, in effect, describe boundaries for a circuit's performance and robustness, and they absolutely must be taken into account for a successful design. Unfortunately, a formal description of the influencing factors and constraints is often missing; it is thus impractical and all but impossible to automate the analog design process, because of the lack of a complete problem definition [SL15, NL09, KMJ13].

Analog circuits also must carry currents that vary widely in magnitude depending on the application: sensor applications require only a few nano-amps, whereas wires in power circuits need to carry up to several amps on a continuous basis. In addition to this broad spectrum of currents, analog circuits may also have to operate reliably at very high temperatures.

The number of nets in analog circuits that are susceptible to electromigration is growing due to technology downscaling and increasingly complex designs. Furthermore, issues arising from Joule heating in conductors must be addressed for these same reasons. Yet another characteristic of analog circuits is that currents in power and signal nets are on the same order of magnitude. As a result, analog designers have been more aware of electromigration issues in recent years than their digital-design counterparts.

A design flow that considers current density in analog ICs is presented in Fig. 3.2. It includes current characterization, current propagation in design hierarchies, current-aware layout PCells, current-aware route planning [LJ03], as well as current-density verification [JL04, JLS04].

The purpose of *current-driven routing* is to ensure that all predefined technology-based current-density constraints are met when the physical layout of the interconnect metallization structures are created. By adopting this "correct-by-construction" approach to obtain an electromigration-robust layout, two closely

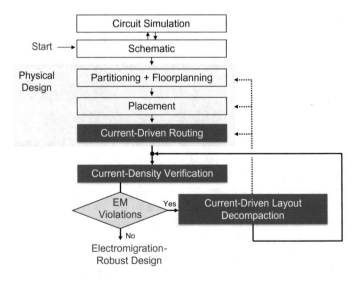

Fig. 3.2 Electromigration-aware analog design flow [Lie06, Lie13]

linked design problems must be simultaneously addressed for automatic and semi-automatic layout generation.

First, it is absolutely necessary to connect to net terminals in a current-density-correct manner to assure the intrinsic reliability of the affected metallization of integrated electronic devices (Sect. 3.5.3). Second, an optimized net topology must be properly planned in order to determine and minimize the current flow within all planned net segments (Sect. 3.3.3). Current-density-correct layout dimensions are then determined based on the current flow within net segments (Sect. 3.5.2). Minimizing the routing area used is a primary optimization goal.

Current-density verification (Sect. 3.5) guarantees the predefined electromigration lifetime of interconnect structures. This is done by verifying that all relevant worst-case current densities within existing metallization structures are always less than or equal to the usage-dependent maximum permitted current-density limits specified for the IC manufacturing technology.

During *current-driven layout decompaction* (Sect. 3.6), an external (de-)compaction tool is used to adjust the previously flagged critical layout structures to their electromigration-correct dimensions. Here, all worst-case current flow values within current-density-critical regions of the net layout are determined based on the results of current-density verification. These current values are then applied to calculate and adapt the correct layout dimensions of all critical layout structures.

To date, commercial analog design tools have required extensive manual intervention to consider the impact of current densities in physical design. Current-density-aware routing tools have been available for some time; they typically widen wires based on terminal currents.

Commercial verification tools for current densities extract a netlist from the layout that includes parasitics. This netlist is then used to simulate the currents in all wires. If any of the resulting current densities exceed an EM-relevant boundary, a violation is detected and highlighted, and an attempt is made to resolve by performing current-driven layout decompaction or earlier steps in the process as necessary, as shown in Fig. 3.2.

3.2.2 Digital Design

One of the main differences between analog and digital circuits is that while analog values and functions are processed in the former, digital values are processed in the latter; because digital values are discrete with respect to time and value, they are less susceptible to disturbances. Furthermore, fewer constraints are needed in the design to assure its proper functioning, which are easily automated. Thus, digital integrated circuits can be designed using well-understood synthesis algorithms; this can be done with *logic synthesis* (which replaces "Circuit Design" in Fig. 3.1).

One can design much more complex circuits in an automated flow than manually. Hence, integrated digital circuits have become increasingly complex with each year, and today can have billions of components and electrical interconnections (nets). At today's level of complexity, they can only be economically designed, verified, and analyzed with the support of computer-aided design (CAD) algorithms. The complexity continues to ramp up rapidly as a result of miniaturization (Moore's Law, Chap. 1) and technological advances, which continue to permit ICs with larger and larger numbers of components to be fabricated.

Automated layout generation, often referred to as *layout synthesis*, is less challenging for digital circuits than with analog circuits, as there are fewer functional constraints. In addition, digital circuits are less sensitive to small voltage changes than their analog counterparts. The proper functioning of digital logic depends essentially on the reliable differentiation between a few different digital logic states. Also, these circuits comprise a relatively small number of different types of subcircuits, so-called gates.

The scalability of the synthesis algorithms (for both logic synthesis and layout synthesis) is key due to the on-going verity of Moore's Law in digital integrated circuits. So-called heuristics are typically used in this context, as they enable the synthesis steps to yield practically "optimal" solutions very rapidly. The layout synthesis of digital circuits can thus be automated with relatively simple algorithms, far more easily than is possible in analog circuit design.

During layout synthesis, verifications should be carried out to check for compliance with predefined constraints, which are based on rules. These verifications are an integral part of the automated design process, because the synthesis algorithms never take all constraints into account and are typically optimized for speed rather than quality—thus the correctness of the output must be verified. Algorithms that solve the given problem in polynomial time should be used in this context, which gives priority to heuristic techniques. The entire circuit cannot normally be fully analyzed, as this would require too much computation time. Instead, the verification process benefits from a number of filtering techniques, which narrow this partially complex task down to a few select portions of the complete circuit, that is, to a few nets, for example. In particular, this is facilitated by investigating the reliability of electrical interconnects, and classifying them into different *net classes*.

There are at least three different net classes in digital circuits. First, there are power distribution networks, where the DC component of the current predominates and currents can be very high. Second, there are clock nets with almost symmetrically alternating currents, where typical root-mean-square (RMS) values can be very high due to the large number of connections to the net. And third, there are signal nets with asymmetrical, pulsed alternating currents, generally with low RMS values.

Nets can be subdivided into these three classes by EM sensitivity, with high (power distribution nets) and low EM sensitivity (clock and signal nets). Verification can thus be restricted to the more EM-sensitive net class(es). However, this simple classification is not absolutely reliable and not all nets can be correctly classified. The verification is only truly reliable if no critical nets are erroneously filtered out, and very few non-critical nets are unnecessarily analyzed. Classification

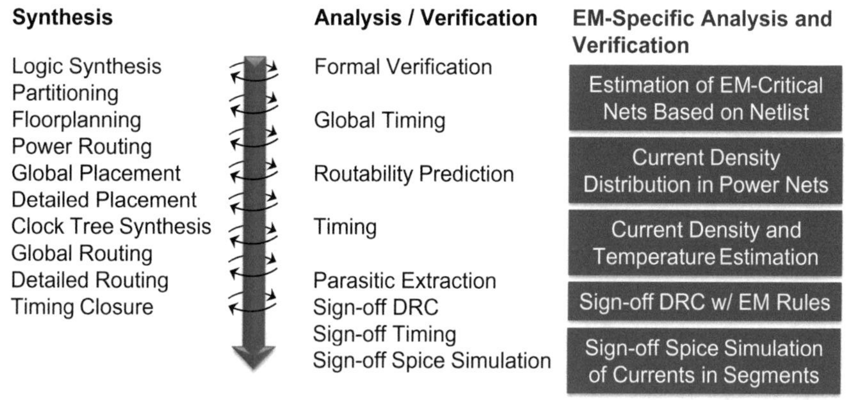

Synthesis	Analysis / Verification	EM-Specific Analysis and Verification
Logic Synthesis	Formal Verification	Estimation of EM-Critical Nets Based on Netlist
Partitioning		
Floorplanning	Global Timing	
Power Routing		Current Density Distribution in Power Nets
Global Placement	Routability Prediction	
Detailed Placement		
Clock Tree Synthesis	Timing	Current Density and Temperature Estimation
Global Routing		
Detailed Routing	Parasitic Extraction	
Timing Closure	Sign-off DRC	Sign-off DRC w/ EM Rules
	Sign-off Timing	
	Sign-off Spice Simulation	Sign-off Spice Simulation of Currents in Segments

Fig. 3.3 Design flow for digital circuits with its typical synthesis-analysis loops [Lie13]. The critical steps—(*left*) synthesis and (*middle*) analysis—are shown, supplemented by options to address current density and other electromigration issues, indicated in the colum on the *right*

and filtering thus benefit from the use of other useful heuristics [JL10], which may require detailed information on the circuit. For example, the maximum currents flowing in every segment must be determined to identify the nets with critical current densities (Sect. 3.3).

A host of synthesis-analysis loops characterizes the design flow for digital circuits. This flow consists of a series of synthesis steps, which methodically concretize circuit geometry (Fig. 3.3, left) [Lie13]. There is a set of verification steps alongside these synthesis steps that ensure the circuit acquires the required electrical characteristics and functions and meets the reliability and manufacturability criteria (Fig. 3.3, middle).

Figure 3.3 (right) shows additional flow options to analyze the impact of electromigration on circuit reliability, which have only been partly supported to date by layout tools. That said, "Sign-off DRC with EM rules" and "Sign-off Spice Simulation" with subsequent current-density verification are now standard functions in state-of-the-art digital layout tools. These functions are also available as stand-alone verification tools.

The first step in these EM-specific analysis and verification options, "Estimation of EM-Critical Nets Based on Netlist," limits the number of nets that must be verified for EM issues [JL10]. It determines the worst-case bounds on segment currents in order to separate signal nets into critical and non-critical sets. Only the set of critical nets, which is typically smaller, requires subsequent special consideration during layout generation.

State-of-the-art commercial design tools typically perform tasks that align closely with the aforementioned analysis steps (see Fig. 3.3, right). They are based on three global current-density limits set to identify EM violations: maximum allowable peak-, average-, and RMS-current densities (Sect. 3.3.1). The actual current-density value in each wire segment is determined by transient or steady state

SPICE (Simulation Program with Integrated Circuit Emphasis) simulations at the transistor level. EM violations are detected if these calculated local current densities exceed a specific limit.

As mentioned previously, power and signal nets are to be verified separately due to their different susceptibilities to electromigration.

3.3 Determination of Currents

EM analysis starts by considering the currents flowing in the interconnects. Transient circuit simulations with SPICE or quasi-static techniques are used to determine the currents in the nets. The resulting maximum, average, and RMS (root-mean-square) values can be converted to an equivalent direct current depending on the effective switching frequency. A current-density value, that can be checked for exceeding a technology, layout, and/or application-specific boundary value, is then determined, in conjunction with the associated layout data.

To identify nets with critical current densities, the maximum currents occurring in each layout segment must be determined. Current vectors are to be created for net nodes that respectively contain the RMS value, the minimum and maximum mean value, and the peak value of the current, as the critical maximum values need not arise in each segment in the same operating mode. The maximum current value can be obtained for each segment from such data. Depending on the frequency, the key dimensioning current is calculated based on these different current values with different weightings.

The EM criticality of a net can be determined by comparing the currents allowed (based on the interconnect cross-section) with the actual currents (Sect. 3.5.2) or by means of current-density simulations (Sects. 3.5.4 and 3.5.5). The latter is based on the calculation of the key current density from the current and available cross-section. Current values are at any rate absolutely necessary.

The different types of damage mechanisms caused by different currents (Sect. 3.3.1), as well as how realistic terminal currents (Sect. 3.3.2) and segment currents (Sect. 3.3.3) are determined, will be dealt with next.

3.3.1 Current Types

The current waveform and the electromigration robustness of an interconnect are closely related. Various studies show an increased electromigration robustness in an interconnect for bi-directional and pulsed-current stress over uni-directional current and constant-current stress [Li89, Ma89, Pi94]. One of the reasons for this is the aforementioned "self-healing effect"—due to a material flow back caused by alternating current directions, which reduces the effective material migration (Sect. 2.4.3, Chap. 2).

When considering wire currents in an electromigration-aware design flow, it is useful to differentiate various current types based on the frequency and characteristics of the current:

- the effective current based on the *root-mean-square value* of the currents (RMS currents) for frequencies lower than 1 Hz,
- the *average-current value* for frequencies greater than 1 Hz, and
- the *peak-current values* to consider for single current pulse events, such as electrostatic discharge (ESD) or single short pulse events in power stages.

The use of these current types has certain limitations and characteristics, as we describe below. In particular, using RMS currents is a more conservative approach, since it does not take into account the self-healing effect. RMS currents are suitable for analog DC nets and reliability-critical applications in general, including long-pulse and continuous-current events.

Dimensioning interconnects based on RMS currents means that thermal effects due to Joule heating in interconnects are considered (which strengthens the above-mentioned "conservative approach"). This so-called self-heating (not to be confused with "self-healing") often limits the maximum current densities, especially in top-level IC metallization layers due to reduced heat conduction.

On the other hand, applying average currents for AC current flows typically considers the self-healing effect of alternating current directions. Average currents are commonly applied to current flows within digital signal nets and analog signal nets. They are also used for EM-lifetime modeling according to Black's Law (Sects. 3.4 and 3.5).

A peak-current flow, such as an individual short-time current flow due to an ESD event, must be considered separately from RMS or average currents. This is due to different damaging effects, such as electrical overstress (EOS), within the metallization requiring appropriate design rules for wire dimensioning.

All three current types should be considered simultaneously when dimensioning interconnect layouts. The current type which leads to the largest number of vias/contacts and the largest wire width in a segment for a given layer and ambient operating temperature must be chosen for layout dimensioning (Sect. 3.5.2).

3.3.2   Terminal Currents

Determining realistic current values for each net terminal (pin) is a problem for any electromigration-aware physical design methodology, notably in analog designs. Most published approaches use a *single* so-called equivalent current value per terminal by considering the current waveform, duty cycle, and frequency.

However, one cannot safely calculate the worst-case current flow in topological connections between layout Steiner points by using only one maximum absolute equivalent current value at a net terminal. (Steiner points are additionally introduced net points to shorten the overall net connection.) This somewhat unexpected

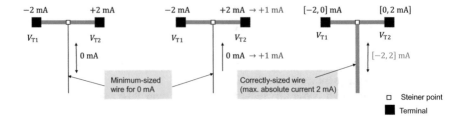

Fig. 3.4 Layout topology of a partial net with average currents [JL10]. A minimum-sized wire is shown at *left* due to a cross-current of 0 mA. Any *change* in the terminal current in terminals V_{T1} or V_{T2} leads to an *increase* in the cross-current that eventually results in a current-density rule violation in the layout of the attached vertical net segment (*middle*). Replacing single terminal currents with current bounds delivers the correct bounds for each net segment (*right*)

statement can be easily understood by considering Fig. 3.4. Suppose we were to perform interconnect wire and via array sizing taking only the maximum absolute equivalent current values at each terminal into account (Fig. 3.4, left). We would then trigger a current-density violation in the Steiner point connecting net segment due to cross-current flows when at least one equivalent current value at a net terminal is subject to change (Fig. 3.4, middle). Hence, a "current value propagation problem" arises within a Steiner point if two connected net terminals are characterized by *reversed* worst-case currents flows.

The lower (average−, RMS−, peak−) and upper (average+, RMS+, peak+) boundaries of the average, RMS and peak equivalent currents at any time, therefore, must be considered if this problem is to be solved [JL10]. These boundaries are derived from the positive and negative parts of the current waveform. The equivalent value for RMS−, which is calculated from the zero and negative parts of the current waveform, must exceptionally be multiplied by a factor of −1.0 to guarantee a lower boundary (RMS− ≤ 0). One can now calculate the cross-currents correctly in any segment of an arbitrary tree-shaped net layout using equivalent current bounds (Fig. 3.4, right).

Next, we introduce three terminal current models, each of which considers the above-mentioned current value propagation problem by utilizing either:

(1) a single current value pair,
(2) a vector of current value pairs, or
(3) a current matrix at each terminal.

In the first terminal current model, which involves a *single current value pair*, the results from one or more simulations are post-processed by calculating a set of *current vectors* satisfying Kirchhoff's current law [Lie06]. They are a snapshot of the circuits operating at the time of minimum and maximum currents *at each terminal* (Fig. 3.5). This reduces the simulation results to a vector of worst-case current value ranges. For a net with m terminals, this may lead up to m current value pairs (i.e., $2m$ current values) attached to each terminal i_{terminal} as follows:

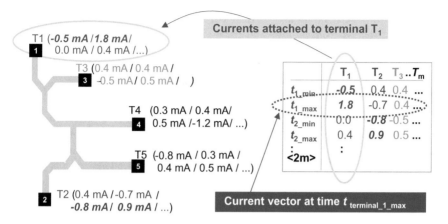

Fig. 3.5 Illustration of the terminal current model utilizing current vectors [Lie06]. Current values assigned to terminals are the respective minimum and maximum values (shown in *italics*) and the current values at the other terminals' minimum and maximum points of time. Every current vector satisfies Kirchhoff's current law, i.e., its current sum is zero

$$i_{\text{terminal}} = \big[\big[i_{i\_\min(\text{terminal}\_1)}, i_{i\_\max(\text{terminal}\_1)} \big],$$
$$\big[i_{i\_\min(\text{terminal}\_2)}, i_{i\_\max(\text{terminal}\_2)} \big], \tag{3.1}$$
$$\cdots$$
$$\big[i_{i\_\min(\text{terminal}\_m)}, i_{i\_\max(\text{terminal}\_m)} \big] \big].$$

Example: $i_{\text{terminal}} = [[-0.5\,\text{mA}, 1.8\,\text{mA}], [0, +0.4\,\text{mA}], \ldots, [-0.2\,\text{mA}, 0]]$

The terminal currents can be obtained by either tracking the current bounds at each terminal during circuit simulation, or by post-processing transient simulation waveforms after the simulation run. The former method is very memory-efficient since no waveforms need to be saved, but it typically slows down the circuit simulation due to additional processing time. The latter post-processing method is more versatile and enables more sophisticated post-processing algorithms to be executed during its processing, for example, to determine the peak current pulse.

All terminal current waveforms must be stored for single current value pair methods—a task which is often impracticable due to the additional storage time required and storage-space limitations. Another disadvantage of this current model is the use of worst-case currents, which produces results that are overengineered, i.e., oversized interconnects, especially if the minimum/maximum current values are only present for a fraction of the circuit's operating time.

A vector with one current value pair for each of n time slots S_x ($x = 1 \ldots n$) is used in the second approach [Lie06]. The minimum and maximum current values of a current value pair are determined between the start and end time of the particular time slot, i.e., a circuit's operating phase. The current values are obtained either by circuit simulation, by manual assignment to the net terminal in the schematic, or by retrieval from a device library. This model accounts for independent current flow events originating from multiple net terminals as follows:

$$i_{\text{terminal}} = [[S_1, i_{\text{min\_1}}, i_{\text{max\_1}}], [S_2, i_{\text{min\_2}}, i_{\text{max\_2}}], \ldots, [S_n, i_{\text{min\_n}}, i_{\text{max\_n}}]]. \quad (3.2)$$

Example: $i_{\text{terminal}} = [[S_1, -1\,\text{mA}, +3\,\text{mA}], [S_2, +2\,\text{mA}, +3\,\text{mA}], \ldots]$

The main advantages of this terminal current model are that (1) the lower and upper RMS, average and peak current bounds can be determined by processing current waveforms; and (2) each time slot can be linked to a circuit operating phase. Advantage (1) yields more realistic functional loads, and advantage (2) enables application mission profiles to be considered (Sect. 3.4). As a result of these two advantages, oversized or undersized interconnects can be avoided.

The third model accounts for *current bounds for a net terminal that are specific to an operating phase* [JL10]. A circuit may have several operating phases according to different operational modes and chip temperatures, such as a normal operating mode, a sleep mode, and a high-temperature operating mode. In this case, the current bounds of a net terminal should be specific to an operating phase and must thus be determined by post-processing the current waveforms obtained for each operating phase.

This model comprises two equally sized current matrices $\bar{L}_n$ and $\bar{U}_n$ (lower bound L and upper bound U). Both matrices hold the equivalent currents for a terminal n ($1 \leq n \leq N$) for each current type o ($1 \leq o \leq O$) at each circuit operating phase p ($1 \leq p \leq P$). The matrix for lower-bound equivalent currents $\bar{L}_n$ is defined as:

$$\bar{L}_n = \begin{pmatrix} i_{Ln,11} & i_{Ln,12} & \cdots & i_{Ln,1P} \\ i_{Ln,21} & i_{Ln,22} & & i_{Ln,2P} \\ & \vdots & & \vdots \\ i_{Ln,O1} & i_{Ln,O2} & \cdots & i_{Ln,OP} \end{pmatrix}. \quad (3.3)$$

Each row in $\bar{L}_n$ represents a P-sized subvector containing the equivalent currents of a specific current type o, and P operating phases. Each O-sized column subvector in $\bar{L}_n$ corresponds to O current types defined for a specific operating phase p (Fig. 3.6). The current matrices of all N instance terminals must have the same dimension.

A matrix for upper-bound currents $\bar{U}_n$ is similarly defined for each terminal n. A set I_n of terminal current matrices with:

$$I_n = \{\bar{L}_n | \bar{U}_n\} \quad (3.4)$$

is then assigned to each terminal of the considered net (see Fig. 3.6, top).

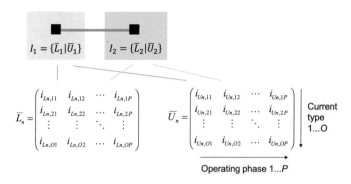

Fig. 3.6 Visualizing two equally-sized current matrices $\bar{L}_n$ and $\bar{U}_n$ for the lower (L) and upper (U) bounds, holding various current types o for two terminals at each circuit operating phase p

The number of current values that have to be stored in $\bar{L}_n$ and $\bar{U}_n$ for a net with N terminals is $(2 \cdot N \cdot O \cdot P)$ regardless of both the number of layout Steiner points in the net and its layout topology. For example, considering a 3-pin net with a pure DC current flow ($O = 3$, $P = 1$), this would result in $2 \cdot N \cdot 3 \cdot 1 = 6\,N$ current values to account for average, RMS, and peak equivalent currents. Sparse storage techniques applied to $\bar{L}_n$ and $\bar{U}_n$ further reduce the computational requirements [JL10].

3.3.3 Segment Currents

Knowledge of segment currents is essential for correct interconnect dimensioning. We can derive the current bounds in net segments using a worst-case current bound approach with one of the terminal current models from Sect. 3.3.2.

We calculate the worst-case current vector $\vec{i}_W$ for each net segment by using the definition of $\bar{L}_n$ and $\bar{U}_n$ in Eq. (3.3), for example. First, each net segment is partitioned into two sets of left-hand side (LHS) and right-hand side (RHS) terminals, which contain the connected LHS and RHS terminals V_{T1} and V_{T2}, respectively (Fig. 3.7, top).

The currents for the LHS and RHS terminal sets are calculated as follows:

$$i_{L_{\text{LHS}}} = \sum_{n=1}^{N_{\text{LHS}}} i_{\bar{L}_{n,op}}, \quad i_{U_{\text{LHS}}} = \sum_{n=1}^{N_{\text{LHS}}} i_{\bar{U}_{n,op}} \tag{3.5a}$$

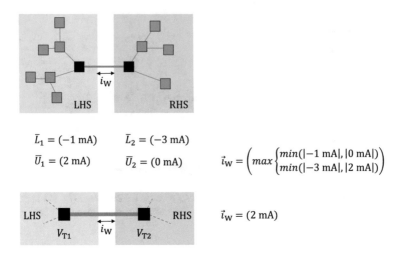

$\bar{L}_1 = (-1 \text{ mA})$ $\bar{L}_2 = (-3 \text{ mA})$

$\bar{U}_1 = (2 \text{ mA})$ $\bar{U}_2 = (0 \text{ mA})$ $\vec{i}_W = \left(max \begin{cases} min(|-1 \text{ mA}|, |0 \text{ mA}|) \\ min(|-3 \text{ mA}|, |2 \text{ mA}|) \end{cases} \right)$

$\vec{i}_W = (2 \text{ mA})$

Fig. 3.7 Determination of the worst-case current value i_w in a net segment between terminals V_{T1} and V_{T2} using the right-hand side (RHS) and left-hand side (LHS) sum of terminal current bounds (with L lower and U upper bound) of a layout topology tree [JL10]. Contrary to Fig. 3.6, this example contains only one current type and one operating phase to simplify the visual presentation

$$i_{L_{RHS}} = \sum_{n=1}^{N_{RHS}} i_{\bar{L}_{n,op}}, \quad i_{U_{RHS}} = \sum_{n=1}^{N_{RHS}} i_{\bar{U}_{n,op}} \tag{3.5b}$$

with $\bar{L}_n$ and $\bar{U}_n$ the corresponding lower- and upper-bound current terminal matrices in the LHS and RHS terminal sets, and N_{LHS} and N_{RHS} the number of terminals in the LHS and RHS sets, respectively [JL10].

An implication of this method is that, for each specific operating phase, terminals will never draw or deliver larger equivalent currents than defined by their current bounds. The entries for the worst-case current vector $\vec{i}_W$ are thus determined for each current type o and operating phase p as follows:

$$i_{W,op} = max \begin{cases} min(|i_{L,LHS,op}|, |i_{U,RHS,op}|) \\ min(|i_{L,RHS,op}|, |i_{U,LHS,op}|) \end{cases}. \tag{3.6}$$

These entries are then used to determine the current-type-specific worst-case currents that are required for sizing wires and via arrays in the net segment between two terminals as follows:

$$i_{W,o} = max(i_{W,op}), \tag{3.7}$$

with $1 \leq o \leq O$, $1 \leq p \leq P$, and terminal indices $1 \leq \{\text{LHS, RHS}\} \leq N$ [JL10].

Figure 3.7 depicts the layout of two terminals V_{T1} and V_{T2}. Here, the current bounds of the left-hand terminal V_{T1} are -1 mA (RMS, average or peak current) for the minimum, i.e., lower bound, and $+2$ mA for the maximum, i.e., upper bound. The lower and upper bounds for the right-hand side terminal V_{T2} are -3 and 0 mA, respectively. Taking the terminal currents at terminals V_{T1} and V_{T2} as a reference, the worst-case absolute current value i_w that can occur in this net segment is 2 mA. The layout of the net segment then must consider this current value and current type for dimensioning.

We note that terminal currents which do not comply with Kirchhoff's current law can also be considered, using the methodology in Fig. 3.7. For example, Fig. 3.8 shows the same approach to determine worst-case bounds in net segments for a net with three different net layout topologies.

It can be shown that all annotated equivalent currents of a defined operating phase and current type can be summated safely. The sum of the annotated equivalent currents of a specific current type is always equal to, or greater than, its equivalent value derived directly from the vector sum of transient current values [JL10].

Having described how to calculate the segment currents, let us now point out a problem that has been omitted thus far: the net topology must be known before segment currents can be calculated. In other words, segment currents can only be calculated *after the entire topology of the net has been laid out*.

This is an issue because current strengths are altered in a previously routed subnet whenever a new terminal is linked to the net. This is illustrated in Fig. 3.9 by connecting Terminal 4 to the net; the currents change depending on the topology of

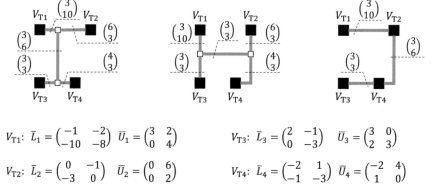

Fig. 3.8 Three layout topologies of the same net with terminals $V_{T1}...V_{T4}$ [JL10]. In each topology, the net segments are labeled with the worst-case current bounds. The terminal current value matrices in this example contain two current vectors (with L lower and U upper bound) representing two different current types (rows), each holding two equivalent current values of an arbitrary current unit (columns)

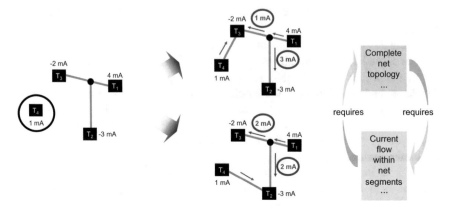

Fig. 3.9 Illustration of the cyclic conflict whereby the net topology, i.e., the sequence of all terminals to be connected, must be known so that currents can be calculated within net segments [Lie06]. At the same time, laying out the sequence of connections requires currents to be known in order to fulfill certain optimization criteria. Single terminal current values are used here for simplicity

this connection. At the same time, however, segment currents should be considered (and hence, should not change) when determining the net topology, that is, when deciding how terminals are connected with each other.

To address this cyclic conflict, a *wire planning* step is required prior to any segment current calculation (Fig. 3.10). A current-driven net topology is determined in wire planning by calculating an optimized routing tree [LJ03]. The net topology is planned and the segment current estimated *concurrently* in this step in order to optimize (minimize) the current flow. Specifically, the major optimization goal of this step should be to minimize the interconnect *area* required for routing, rather than simply minimizing the length. In other words, current-intensive segments are kept as short as possible (Fig. 3.10, right).

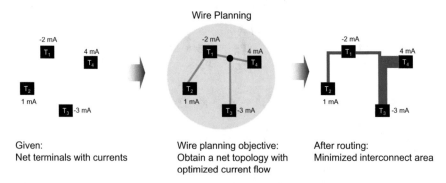

Fig. 3.10 Solving the cyclic conflict in Fig. 3.9 by current-driven wire planning. This additional step in an electromigration-aware design flow creates a list of global connections and Steiner points to optimize (minimize) current flow, thereby minimizing the interconnect area after routing

After the net topology is defined, segment currents can be calculated (see above) which are then used to obtain the correct layout sizes for wires and via arrays (Sect. 3.5.2). Any detailed routing is subsequently simplified to a point-to-point routing between net terminals with *known* segment widths.

3.4 Determination of Current-Density Limits

Having determined realistic current values in the terminals and segments, we now turn our attention to the resulting current density and its limitations, i.e., its boundary values.

Verified current-density limits for an interconnect serve as a major constraint on any electromigration-aware design flow. While we may be tempted to consider current-density limits as "static" properties, it is important to realize that the application for which the IC will be used (e.g., automotive applications that involve high temperatures, as compared to consumer devices designed to operate at room temperature) will affect such current-density limits. Indeed, all of the relevant environmental conditions and functional loads that each electronic system, module and component must be able to sustain during manufacturing, assembly, storage, and operation needs to be considered to determine robust application-specific current-density limits. This is done by formalizing these loads and conditions in so-called mission profiles, which specify environmental (e.g., temperature) and other conditions to which the chip will typically be subject when used for that mission (e.g., used in an automotive engine compartment).

3.4.1 Mission Profile

A *mission profile* comprises all information about environmental conditions and functional loads, such as I/O currents, of a module or component. By definition, any mission profile is specific to a given electronic circuit; however, they are often generalized into standardized mission profiles that are used to account for various application classes, such as consumer, automotive, industry, medical, and aerospace. This standardization improves their applicability.

The operating-state definitions of the mission profile in combination with the specification requirements are used to derive a set of simulation stimuli for each operating state [JK14]. These stimuli are subsequently used to obtain all effective terminal currents within the electronic circuit for each operating state (Sect. 3.3.2). These currents are then applied during the IC physical design phase to lay out all interconnects with the correct cross-sectional areas, and to connect to all terminal pins without causing current-density violations (Sect. 3.5.2). Compliance with these electromigration design rules is verified in the final layout result, using current-density verification techniques (Sect. 3.5).

When a mission profile is applied to an existing, validated physical layout (e.g., applied to a layout characterized by "reference" parameters), the physical layout must be validated once again with respect to current density. That is, a physical design validation with respect to current density is required every time the temperature profile or the duration of operating phases are modified, despite maintaining the same physical layout. This is because the application of the mission profile yields new current-density design rules, which differ from the reference design rules in the design rule manual, and that must be obeyed by the physical design as a condition of validation. Hence, the validation of current-density limits is crucial for the reuse of physical layout.

How do we obtain application-robust design rules, such as current-density limits? First, the fundamental parameters of the electromigration failure model of a technology, such as the activation energy, are determined during technology reliability *characterization*. (Parameters applied and obtained during this step are subsequently indicated with the subscript "char".) Accelerated test conditions, such as obtaining current-density limits j_{char} under increased temperature T_{char}, are in place here. Second, environmental conditions are scaled to so-called *reference* conditions to make design rules applicable to designers. For example, a certain current-density limit that is valid for a specific temperature is given to an IC designer. Hence, reference conditions (subscript "ref") provide a generic frame for a wide range of IC applications. Finally, a particular IC application is defined by *effective* environmental conditions in which it is used (subscript "eff"). For example, the temperature T_{eff} and the current density j_{eff} depend directly on the application environment and usage, that is, the specific mission profile (Fig. 3.11).

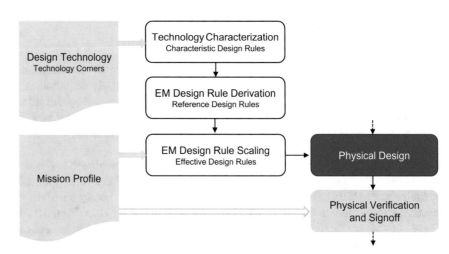

Fig. 3.11 Applying technological constraints and a mission profile to obtain characteristic design rules, reference design rules, and effective design rules [JK14]. The resulting application-robust current-density limits are applied during physical design and verification

In the next Sect. 3.4.2, we describe these steps in more detail, up to and including the final calculation of the application-robust current-density limit, which is represented by the mission-profile-specific effective current density j_{eff} for each interconnect layer.

3.4.2   Application-Robust Current-Density Limits

As we introduced in Sect. 2.2 (Chap. 2), electromigration median time to failure (MTF) can be modeled according to Black's Law [Bl69a] as follows:

$$\text{MTF} = t_{50} = \frac{A}{j^n} \cdot \exp\left(\frac{E_a}{k \cdot T}\right), \tag{3.8}$$

with t_{50} as the time when 50% of all interconnects have failed, j as current density in the interconnect metallization, E_a as layer-specific electromigration activation energy, n as dimensionless current-density scaling factor, k as the Boltzmann constant, T as interconnect temperature, and A as a cross-section-dependent constant, that, among others things, relates the rate of mass transport to median time to failure (MTF) [Bl69b].

The terms E_a, A, and n are all closely associated with the specific technology being used, and values for these terms are established during technology characterization [JP001, JESD61]. These values must be monitored closely during manufacturing, since the physical layout has been constructed for correctness with respect to these characterization values [JEP119A, JESD61]. The environmental stress conditions defined in the mission profile are reflected in the temperature term T. The term j is used to represent the functional load (applied uniformly in the wire cross-section area), that is, the electrical current i (see Eq. 3.7) as well as for the physical design solution, namely the cross-section area A_{wire} of the interconnect, as follows:

$$j = \frac{i}{A_{\text{wire}}}. \tag{3.9}$$

The entire circuit may fail if even a single interconnect is damaged; thus, Eq. (3.8) is not suitable to describe the electromigration reliability target for an entire IC, but will require adjustment. In particular, we rewrite Eq. (3.8) as follows, to describe and define the actual component-specific EM design rule targets:

$$t_{\text{life,ref}} = \text{AF}_{\text{TF}} \cdot \text{AF}_{\text{T}} \cdot \text{AF}_{\text{q}} \cdot t_{50,\text{char}} = \text{AF}_{\text{TF}} \cdot \text{AF}_{\text{T}} \cdot \text{AF}_{\text{q}} \cdot \frac{A}{j_{\text{char}}^n} \cdot \exp\left(\frac{E_a}{k \cdot T_{\text{char}}}\right),$$

$$\tag{3.10}$$

with the three *application scaling factors:*

- target application lifetime AF_{TF},
- effective interconnect temperature AF_T, and
- permissible failure rate at the end of life AF_q [JK14].

The scaling factor of target application lifetime AF_{TF} is used to tailor the maximum current-density limits in interconnects to specific application requirements and optimization goals, such as long interconnect lifetimes. For instance, the interconnect lifetime for ICs used in many applications is set to $t_{life,ref} = 10$ years of constant operation at a constant reference temperature T_{ref} (e.g., 378 K [105 °C]). If the IC application is to be used for $t_{life,ref} = 15$ years instead, then AF_{TF} is set to 1.5 (=15y/10y).

The two remaining scaling factors, of effective interconnect temperature AF_T and of permissible failure rate AF_q, are obtained as follows [JK14]:

$$AF_T(T_{ref}, T_{char}) = \exp\left[\frac{E_a}{n \cdot k} \cdot \left(\frac{1}{T_{ref}} - \frac{1}{T_{char}}\right)\right], \qquad (3.11)$$

$$AF_q(0.5, q_{ref}) = \frac{S_1}{\exp[\mathrm{norminv}(q_{ref}) \cdot \sigma]}. \qquad (3.12)$$

We refer to the term $t_{life,ref}$ (Eq. 3.10) as the *targeted* reference lifetime, which represents the time at which the cumulative fraction, i.e., a quantile q_{ref}, of interconnects is projected to fail while being subjected to a constant reference temperature T_{ref} and stressed with a reference current density j_{ref}. We use the term $t_{50,char}$ to denote the time when 50% of interconnects have failed during characterization, at a temperature T_{char}, while being stressed with a current density of j_{char}. In Eq. (3.12), we use σ to represent the lognormal standard deviation, which is obtained from technology characterization. The variable S_1 is used to account for the low confidence level during technology characterization; in practice, values for S_1 in the range of $1 < S_1 \leq 10$ are used [JK14].

Figure 3.12 visualizes the application scaling factors introduced above. The green solid line represents the measurements of the cumulative failure distribution $F(t)$ for electromigration taken at reference conditions during the initial technology reliability characterization. This characterization is done with time acceleration by using high temperatures (typically > 473 K [200 °C]) and high current densities. As a result of this "time acceleration," the green line is the left-most one, expressing shortest lifetimes (*x*-axis).

The factors AF_{TF}, AF_T, and AF_q represent the independent scaling factors to either (i) determine the maximum effective lifetime of the application (cf. Eq. 3.19) or (ii) to calculate the maximum permitted current-density limits to be obeyed during IC layout design to guarantee a minimum effective lifetime (cf. Eq. 3.20). Specifically, the target application lifetime factor AF_{TF} is used to tailor the maximum current-density limits in interconnects to specific application requirements and optimization goals, such as long interconnect lifetimes. Furthermore, the

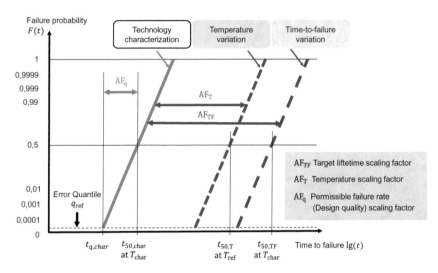

Fig. 3.12 Visualizing the application scaling factors [JK14]. AF_{TF} accounts for the effective target lifetime adjustment compared to reference conditions, AF_T for the ambient temperature profile adjustment and AF_q for design-specific aspects, such as the number of interconnects. Based on known characterization values taken at reference conditions (green line), these factors are used to either calculate the maximum effective lifetime of the application, or to determine the maximum permitted current-density limits when designing the chip

temperature factor AF_T is applied to adjust the maximum current-density limits to different chip operating temperatures, which are defined in the mission profile. The factor AF_q accounts for design-specific aspects, such as the number of interconnects.

In Fig. 3.12, AF_T is illustrated using temperature variation that represents a temperature *decrease*, as compared to the technology characterization temperature (which was significantly higher, to speed up the characterization process). Thus, we can see that AF_T has indeed extended the time to failure, as illustrated by the shifting to the right of the blue dashed temperature variation line, as compared to the green technology characterization line.

The parameter $t_{\text{life,ref}}$ (Eq. 3.10) is considered as target electromigration lifetime, from which j_{ref} is finally derived. This target lifetime of a circuit should not be confused with its cumulative electromigration operating lifetime $t_{\text{life,eff}}$ that accumulates the IC operating times at specific temperature levels. Obviously, the IC's (actually accumulated) effective operating lifetime $t_{\text{life,eff}}$ must be shorter than the (verified) target lifetime $t_{\text{life,ref}}$ in order to meet the validation criteria.

Clearly, $t_{\text{life,ref}}$ must scale with rising reliability requirements (Eqs. 3.9–3.11). Further, the larger the number of interconnects on an IC, the smaller the error quantile q_{ref} needs to be chosen (Eq. 3.12). This then results in a higher electromigration target lifetime $t_{\text{life,ref}}$.

As noted above, the variable S_1 in Eq. (3.12) accounts for the low confidence level during technology characterization. The low level of confidence is a statistical

result of only having been able to efficiently characterize a few hundred interconnect structures. To remedy this, the characterization result is scaled in subsequent
applications to a much larger number, representing millions or even billions of
interconnects. As a result, S_1 must be set to values greater than 1.0 ($1 < S_1 \leq 10$)
[JK14]. Additionally, we can see from Eq. (3.11) that temperature scaling can
significantly impact application reliability, and thus must be carefully considered
during the layout implementation and verification processes.

Finally, the layer-dependent reference current density j_{ref} at T_{ref} is derived from
Eqs. (3.9)–(3.11) as follows:

$$j_{ref}(T_{ref}, q_{ref}) = \frac{j_{char}(T_{char})}{AF_T(T_{ref}, T_{char}) \cdot AF_q(0.5, q_{ref}) \cdot AF_{TF}}. \tag{3.13}$$

Recall that a particular IC application is defined by the *effective* environmental
conditions in which it is used. Hence, for most applications, the reference electromigration design rules must be scaled to effective, i.e., application-specific
design rules. This accounts for usage conditions as defined by the mission profile as
follows:

$$t_{life,ref}(q_{ref}) \rightarrow t_{life,eff}(q_{eff}), \tag{3.14}$$

$$T_{ref} \rightarrow T_{eff}, \tag{3.15}$$

$$j_{ref} \rightarrow j_{eff}. \tag{3.16}$$

The application-specific effective error quantile q_{eff} is calculated as:

$$q_{eff} \ll \frac{1}{m}, \tag{3.17}$$

where m is the number of interconnect segments routed on the circuit.

The complexity of the IC and the electromigration reliability budget are captured
in the term q_{eff}. For a given mission profile, we use T_{eff} to denote the effective,
application-specific temperature at which an arbitrary current flow would cause the
same cumulative electromigration damage that it would at T_{ref} with the reference
mission profile. (In this approach, we make the conservative and reasonable
assumption that the current flow itself causes no significant self-heating in the
interconnects.) One can calculate T_{eff} as follows [JK14]:

$$\frac{1}{T_{eff}} = \frac{1}{T_{ref}} - \frac{k}{E_a} \cdot \ln\left(\frac{t_{life,eff}}{t_{life,ref}}\right), \tag{3.18}$$

with:

$$t_{life,eff} = \sum_{s=1}^{S}\left[t_{life,s} \cdot \exp\left[\frac{E_a}{n \cdot k} \cdot \left(\frac{1}{T_{ref}} - \frac{1}{T_s}\right)\right]\right], \tag{3.19}$$

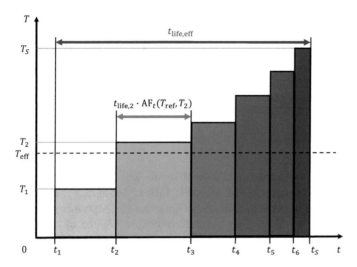

Fig. 3.13 Temperature versus cumulated effective lifetime profile [JK14]. This plot shows an example of cumulated time durations an IC has to sustain at various temperatures according to its mission profile

where T_s is the average temperature at circuit operating state s, $t_{\text{life},s}$ is wall clock duration of s, $t_{\text{life,eff}}$ is the effective electromigration lifetime at T_{eff}, n is the dimensionless current-density scaling factor from Eq. (3.8), and $t_{\text{life,ref}}$ is obtained from Eq. (3.10).

An example of this relationship between IC operating temperatures T_x applied during a time interval $[t_{x+1} - t_x]$ and the effective electromigration lifetime $t_{\text{life,eff}}$ is depicted in Fig. 3.13. The time intervals on the x-axis represent the accumulated time interconnects spend at a certain temperature T_x. For practical applications, T_x is often given as mid-temperature of a temperature interval itself to allow some temperature categorization in mission profiles (e.g., 100 h @ $T_x = 373$ K [100 °C] actually represents 100 h @ $T_x = 368$–378 K [95–105 °C]).

We have seen how T_{eff} depends on the application mission profile; it is important to note that T_{eff} is also influenced by the interconnect layer through the terms n and E_a in Eqs. (3.10), (3.11), and (3.18). As a result, it is important to calculate T_{eff} separately for each interconnect layer of the applied technology, and furthermore to use the highest value obtained as the reference for physical layout as well as for current-density verification.

By reusing Eqs. (3.10)–(3.12), the mission-profile-specific effective current-density limit j_{eff} for each interconnect layer can now be calculated as follows [JK14]:

$$j_{\text{eff}}(T_{\text{eff}}, q_{\text{eff}}) = \frac{j_{\text{ref}}(T_{\text{ref}})}{\text{AF}_T(T_{\text{ref}}, T_{\text{eff}}) \cdot \text{AF}_q(q_{\text{ref}}, q_{\text{eff}}) \cdot \text{AF}_{\text{TF}}}. \tag{3.20}$$

In summary, application-robust current-density limits are obtained by considering
(i) the technology reliability characterization, (ii) design-specific properties, such as
the number of nets in an IC and the quality level to be achieved, and (iii) the
specific mission profile of the IC application, such as the temperatures at which the
IC must operate, and the duration at such temperatures.

3.5   Current-Density Verification

Current-density verification validates that all given current-density constraints
within the interconnect layout are met. Given the aforementioned application
conditions for an IC, such as temperatures, a successfully performed current-density
verification assures the predefined electromigration lifetime of interconnect structures. This is done by verifying that all relevant worst-case current densities within
metallization structures are always smaller or equal to the application-robust
current-density limits as calculated in Sect. 3.4.

Figure 3.14 visualizes an excerpt of the graphical output of a current-density
verification tool.

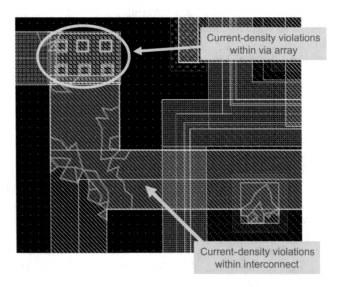

Fig. 3.14 Excerpt of a current-density verification layout with flagged violations marked in red
[JL04]. These violations must be subsequently addressed by adjusting the wire widths and
extending the via array

3.5.1 Methodology

Current-density verification methods require the following input data:

- a set of current values as boundary values (Sect. 3.3),
- an appropriate representation of the layout geometry,
- technology-dependent data (e.g., layer thickness),
- temperature data (e.g., average chip temperature or a temperature field plot), and
- the maximum permitted current (density) in each layer (Sect. 3.4).

A plot of thermal simulation data is often used to account for temperature gradients near significant heat sinks and heat sources.

A number of current-density-verification solutions are available, which address different complexities and constraints in analog and digital design. In one widely used approach, worst-case current flow values within each net segment are extracted and the required technology-dependent interconnect cross-section area is then determined (Sect. 3.5.2). The required, and available, cross-section areas of each net segment and via are then compared during the verification step, which is often part of the regular layout verification suite. This approach is suitable for the verification of pure digital layouts and can be used for fast pre-verification of complex analog and mixed-signal layouts.

Another, more current- and net-specific approach uses FEM-based current-density modeling and calculation for verification (Sects. 3.5.4 and 3.5.5). Here, the calculated current densities within interconnect structures are compared with the maximum current-density limits. An FEM-based verification typically comprises four basic steps [JL04]:

(1) current-density verification of net terminals (intrinsic reliability),
(2) determination of non-critical nets, and their de-selection from further verification,
(3) calculation of current densities within the given metallization layout and comparison with layer-dependent current-density limits, and
(4) evaluation of the violations obtained.

We next discuss these four steps in further detail. First, a current-density verification of net terminals is performed to ensure that the metallization of the net terminals sustains the assigned current values.

Next, non-critical nets should be excluded from further checking, so the verification process may complete more quickly. The criticality of a net is determined by considering the sum of the worst-case current values of each net terminal (without accessing the net's actual layout topology). The net is excluded if this sum is smaller than the maximum permitted current on the minimum-sized metallization layer. Recall that the maximum permitted currents per layer can be easily derived from any given current-density limit as the layer-specific interconnect dimensions are known.

The layout topology of critical nets is then extracted and the worst-case equivalent and ESD-current-flow conditions are determined by calculating all possible up- and down-stream current strengths. This methodology makes it possible to cut the net layout into smaller and independent segments. The determined worst-case currents are then assigned to the segment cuts. All given worst-case currents can thus be verified simultaneously with only one detailed current-density calculation cycle per independent layout segment.

The detailed current density within the metallization of a layout segment is determined by using the finite element method (FEM). Here, the layout is further segmented into finite elements (e.g., triangles) and the current density is calculated using the potential field gradient. The current density is then compared with its maximum permissible value within each finite element.

Finally, after the violating finite elements, i.e., layout regions, have been identified, the verification results are analyzed to distinguish "dummy errors," such as current-density spots at corner coordinates, from real violations.

An FEM-based verification approach is very precise and thus suitable for current-density verification within arbitrarily shaped digital, analog, and mixed-signal layouts. It has also been successfully applied to current-density verification in via structures [JL04].

The following subsections describe various aspects that must be considered when implementing current-density verification and using it during physical design. First, we outline specific equations that can be used to determine the appropriate wire and via sizes (Sect. 3.5.2). Next, an often overlooked issue, the intrinsic reliability of net terminals is introduced (Sect. 3.5.3). Finally, we describe simulation methodologies that can be applied for EM (Sects. 3.5.4 and 3.5.5). That is, we will be discussing how the finite element model is represented for current-density simulation.

3.5.2 Current-Required Wire and Via Sizes

Equations (3.21)–(3.23) (first published in [LJ05]) have been shown to be most accurate for calculating the nominal wire width $w_{\text{nom}}(T_{\text{eff}})$. They are based on the maximum permitted (application robust) current-density limits $j_{\text{eff,eq}}$ and $j_{\text{eff,peak}}$ determined for a specific application temperature T_{eff} according to the mission profile (Sect. 3.4), as follows:

$$w_{\text{nom}}(T_{\text{eff}}) = \max \begin{cases} \dfrac{i_{\text{w,eq}}}{j_{\text{eff,eq}}(T_{\text{eff}}) \cdot h_{\text{nom}}}, & (3.21) \\[3ex] \dfrac{i_{\text{w,peak}}}{j_{\text{eff,peak}}(T_{\text{eff}}) \cdot h_{\text{nom}}}, & (3.22) \\[3ex] w_{\text{min\_process}}. & (3.23) \end{cases}$$

These equations also include the nominal layer height h_{nom}, a process-dependent minimal wire width $w_{\text{min\_process}}$, and the equivalent (RMS or average) currents $i_{\text{w,eq}}$

and peak currents $i_{w,peak}$. These currents represent the worst-case absolute current value i_w of segments as obtained in Sect. 3.3.3 (derived from the terminal current models in Sect. 3.3.2).

When determining an effective (application-robust) wire width $w_{eff}(T_{eff})$, technology characteristics, such as the ratio between nominal and minimum layer height (h_{nom}/h_{min}), wire width variation Δw, and etch loss w_{etch}, also must be considered, as follows:

$$w_{eff}(T_{eff}) = w_{nom}(T_{eff}) \cdot \frac{h_{nom}}{h_{min}} + \Delta w + w_{etch}. \tag{3.24}$$

In addition to adjusting the wire widths, vias also must be sized according to their currents. Vias are normally adjusted for current (density) by replacing a single via with an array of vias, or altering the number of vias in a via array. The current- and temperature-dependent number of vias $n_{via}(T_{eff})$ required within a via array is determined as follows:

$$n_{via}(T_{eff}) = \text{ceil}\left(\frac{i_{w,eq}}{i_{single_{via}}(T_{ref})} \cdot f(T_{eff}) \cdot g(H) \right), \tag{3.25}$$

where $i_{w,eq}$ is the worst-case equivalent current the via array must sustain, i_{single_via} characterizes the maximum permissible current value of a single via at reference temperature T_{ref}, and $f(T_{eff})$ is obtained as in Eq. (3.26) [LJ05].

The factor $g(H)$ accounts for the inhomogeneity of the current flow. The term $g(H)$ is set to $g(H) = 1.0$ for homogeneous current flow, i.e., an equally distributed current density throughout the wire cross-section, otherwise, it is set to $g(H) > 1.0$ for inhomogeneous current flow, according to FEM simulations in [LJ05].

In many cases, the maximum permissible current value that a single via can sustain is only known for a specific reference temperature. In order to achieve more flexibility, a temperature scaling factor can be derived from Eq. (3.8) which takes any temperature-related difference of this maximum current into account. This temperature scaling factor $f(T_{eff})$ is calculated as follows:

$$f(T_{eff}) = \exp\left(-\frac{E_a}{n \cdot k \cdot T_{ref}} \left(1 - \frac{T_{ref}}{T_{eff}} \right) \right), \tag{3.26}$$

where E_a is the activation energy of the electromigration failure process in the via material, k is the Boltzmann constant, T_{eff} is the effective application temperature, and n a scaling factor (usually $1 \leq n \leq 2$) [LJ05].

Please note that the term $f(T_{eff})$ [and, hence, Eq. (3.26)] is only of significance if the actual (effective) application temperature T_{eff} differs from the reference temperature T_{ref} that has been used to determine the maximum current per via. Otherwise, $f(T_{eff}) = 1$ in Eq. (3.25).

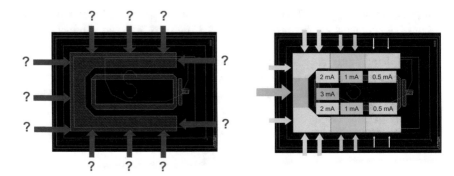

Fig. 3.15 Terminal regions of a U-shaped analog pin consisting of one metallization layer. While such pins can be connected from different angles (left), designers must keep in mind that terminal regions must be verified with regard to their maximum permissible currents (right). These values must be considered when connecting the wire(s) to the terminal

3.5.3   Net Terminal Connections

Analog device terminals (pins) are distinguished by their great variety of shapes and sizes. When connecting such a terminal to a wire (interconnect), designers must bear in mind that different connection positions of a wire to the terminal can produce different current loads within the terminal structure. A current-density verification should therefore include not only the interconnects but also all terminal structures.

While designing the interconnects (the routing step), it is advisable to determine the ampacities, i.e., the current carrying capacities, of the various terminal regions (Fig. 3.15) and compare them with the maximum current(s) in the wire(s) reaching the terminal. The terminal areas of ampacity below the expected maximum wire current should then be eliminated as potential connecting points.

3.5.4   Simulation Methods for Electromigration Processes

Implementing a current-density verification tool requires the *simulation* of current density, as the latter cannot be measured directly within the IC's interconnect. Let us therefore first consider the various simulation methods that can be applied.

Electromigration is a stochastic process, and as such, results obtained from (deterministic) simulations should therefore be treated as random variables. The simulation results can only be used successfully in the design of electronic circuit layouts when they are combined with variables that describe their distribution. Methods based on the statistical analysis of the results should therefore be used for modeling, simulating and assessing EM phenomena. An alternative option is to apply safety factors in the analysis to assure the robustness with a specific probability.

As already mentioned in Chap. 2, migration is a complex problem that can be described by a system of differential equations. Several solving strategies exist for this type of mathematical problem. The related simulation methods can be classified as follows:

- analytical methods,
- quasi-continuous methods,
- concentrated or lumped element methods, and
- meshed geometry methods, such as

  - finite element method (FEM),
  - finite volume method (FVM), and
  - finite differences method (FDM).

When using these methods, it is important to respect the numerous boundary conditions given by the simulation problem, as well as the coupling effects of the different physics domains involved in the migration process. The solution space consists of a set of variables (also referred to as "degrees of freedom") that must be adjusted to fit the boundary conditions and equations.

For a detailed description of the above-mentioned first three simulation methods please refer to Sect. 2.6.1 in Chap. 2. We re-iterate below the fourth approach, meshed geometry methods for EM analysis.

Meshed geometry methods offer several advantages for migration analysis. The degrees of freedom can be spatially resolved in a variable manner by adjusting the mesh granularity. The calculation effort is limited due to the bounded degrees of freedom—the mesh is finite. Using only basic geometries for the mesh elements further simplifies the simulation.

The *finite element method* (FEM) is a universal tool for calculating elliptic and parabolic equation systems. It is a numerically very robust method. Many tools support FEM due to its great variety of applications. The system of equations is built from degrees of freedom for nodes and elements.

The *finite volume method* (FVM) uses polyhedrons to divide the given geometry, while solving the equations only at the center of each polyhedron. FVM is best suited for conservational equations, such as mass flow calculations for fluid and gas transport. It could be applied to migration when modeling atomic flux similar to gas diffusion.

The *finite differences method* (FDM) is numerically very simple and therefore well suited for theoretical analysis or very fast calculations. Due to its simplicity, its results are not as exact as with FEM. As its name suggests, the system of equations is based on the differences in the degrees of freedom.

The finite element (FE) and the finite difference methods (FDMs) are particularly well suited for the diffusion or heat equation [Eq. (2.2), Chap. 2] that is often applied to model EM.

Modern FEM tools enable a multi-physics approach for different abstraction levels [Br04, El16]. (Multi-physics is a computational discipline which uses simulations that involve multiple physical models or multiple simultaneous physical phenomena.) This multi-physics approach is very helpful, especially in calculating the temperature gradients or mechanical stresses in a layout.

In the following Sect. 3.5.5, we limit our description of current-density simulation to the use of FEM, as it is the most convenient option for fast EM analysis.

3.5.5   Current-Density Simulation

Numerical methods (e.g., FEM) are necessary for calculating the current density for many geometrical configurations. While in some cases, two-dimensional (2D) models can be used, three-dimensional (3D) models are typically required to produce results with the required fidelity. Quasi-3D or 2.5D models have been investigated and deployed in the past [JL04]. However, sufficiently precise simulation results can only be achieved for relatively coarse wiring structures with these halfway-house approaches.

In an effort to limit the size of models and thus computation times, symmetries can be used so that only a half, a quarter or even an eighth of the entire lattice structure must be modeled. The (wiring) structures in an IC layout can be halved under the best circumstances, as the critical structures of interest are asymmetrical within the layer or are vertical via interconnects.

The diffusion or heat equation [Eq. (2.2), Chap. 2] for calculating the current density is employed as the mathematical model. Here, the boundary conditions are an electrical potential and a current density, each at the boundary surfaces in the model (Fig. 3.16).

The internal variables (degrees of freedom) of the nodes or elements are the electrical potential and the current density. The calculation is based on Ohm's Law as follows:

$$\vec{j} = \sigma \cdot \vec{E}. \tag{3.27}$$

Fig. 3.16 Model for current-density calculations with boundary conditions at the boundary surfaces (electrical potential $\Phi_1 = 0$ and current density $\vec{j_2} = j_0 \cdot \vec{n}$)

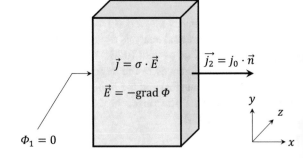

$$\vec{j} = \sigma \cdot \vec{E}$$
$$\vec{E} = -\text{grad } \Phi$$

$$\vec{j_2} = j_0 \cdot \vec{n}$$

$$\Phi_1 = 0$$

The current density is $\vec{j}$, σ is the electrical conductivity, and $\vec{E}$ the electrical field strength, that is:

$$\vec{E} = -\mathrm{grad}\,\Phi, \tag{3.28}$$

where Φ is the electrical potential.

The set of equations to be solved is a system of first order, i.e., linear differential equations with a sparse matrix, as only neighboring elements in the network model are related. This simplifies the computational complexity considerably.

The current density and other driving forces of interest, such as the temperature gradient and the mechanical stress gradient, can be determined with a static solution of the set of equations, whereby a new set of equations with different boundary conditions must be solved for every driving force. This involves calculating the current density, the thermal conduction, and the mechanical stress (Sect. 2.6.3).

EM and thermal and stress migration are closely coupled processes, whose driving forces are linked with each other and with the resultant migration change (Sect. 2.5.3). Due to this coupling, the simulation of the atomic flux for all migration types (Sect. 2.6.2) can become a time-consuming task.

If void creation over time is to be simulated, as well, an iterative, transient process is required, and the model must be modified in each iteration step (Sect. 2.6.4) [WDY03, WCC+05].

For a fast simulation, the calculation of the driving forces only is to be performed. The results of this simulation are the current density, temperature, and mechanical stress fields of the interconnect. These fields indicate locations with a high risk of void creation. The pace of void and hillock growth can be estimated by deriving the atomic diffusion flow based on these three fields. The growth velocity, in turn, determines the time to failure of the interconnect. This necessitates, however, that the critical void volume or an equivalent parameter restricting time to failure must be known (Chap. 4).

3.6 Post-Verification Layout Adjustment

Once current-density verification has been performed, the flagged wires and vias with current-density violations must be modified in order to produce an electromigration-robust layout design. Current-driven layout decompaction has been shown to be an effective point tool to avoid repeated place and route cycles when addressing current-density verification errors. Its major goals are the post-route adjustment of layout segments according to their actual current density, and a homogenization of the current flow.

Prior to layout adjustment, the current-density verification tool must identify regions with excessive current-density stress. Based on these data, four steps are performed (Fig. 3.17) [JLS04]:

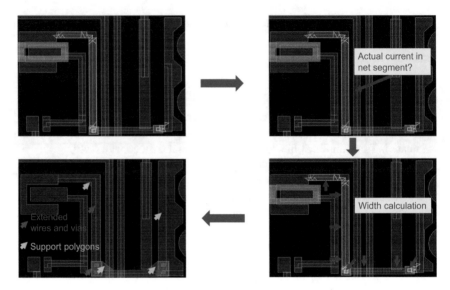

Fig. 3.17 Illustration of current-driven layout adjustment in analog design. For any net segment and via with a current-density violation (upper left, red), the actual current is calculated (upper right) and the appropriate segment width and via size/number subsequently calculated (lower right). After shifting neighboring elements (i.e., layout decompaction), wires and vias are extended and support polygons are implemented (lower left)

(1) layout decomposition,
(2) wire and via array sizing,
(3) addition of support polygons, and
(4) layout decompaction.

First, all net segments are retrieved from the given net layout during layout decomposition. The end points of each segment (i.e., net terminals or layout Steiner points) then represent either (artificial) current sources or current sinks.

Next, the current within a net segment is determined using the location-dependent current-density data obtained from the prior current-density calculation. The appropriate cross-section areas of critical wires and via arrays (Sect. 3.5.2) are then calculated based on these current values.

Finally, so-called support polygons are added to critical layout corners (e.g., wire bends) and around net terminals to reduce the local current-density stress if wire widening is not applicable (e.g., at terminals) or sufficient (e.g., addressing current-density spots in corner bends) [JLS04].

The final layout decompaction with cross-section area adjustment can be performed with any layout decompaction tool capable of simultaneous compaction and decompaction of layout structures while preserving the net topology.

3.7 Design Options for Electromigration Avoidance

So far we have shown how to address electromigration issues during the physical design steps of an electronic circuit. While these are effective means of adjusting the physical layout to the currents (and thus current density) in today's design flows, we would like to next discuss electromigration avoidance in a broader sense. These farther-reaching solutions are subsequently investigated in detail in Chap. 4.

Clearly, five different options come to mind that allow circuit designers to combat electromigration:

(1) limiting the current density,
(2) limiting the temperature,
(3) performing layout modifications to take advantage of EM-inhibiting effects,
(4) modifying materials with classical technologies to eliminate EM damage mechanisms, and
(5) selecting new, EM-resistant materials.

The first mentioned *current-density limitation* requires either restricting the current or increasing the interconnect cross-section. The current can be reduced by limiting the operating frequency. However, this would have a negative effect on the required performance gain in modern ICs. Applying the second option, i.e., increasing the cross-sectional area, asks for a greater interconnect width. However, it is not only impossible to widen all wires on a chip, it would also be counter-productive to any new technology whose benefit is derived from its reduced structure size.

Both measures for restricting the current density are thus impracticable, as they work against the general trend in technological advancement.

The *temperature* can only be influential to a limited extent. The reason for this is that possible local structural changes in the layout design only impact temperature gradients, as the critical ambient temperatures cannot be changed by the designer. Heat exchange can, for example, be improved and thus the temperature gradient reduced by inserting additional metal structures, in the form of thermal wires or thermal vias, for example.[1] This type of measure is commonly used in three-dimensional integrated circuits [LM06].

It is also worth noting that the temperature gradients occurring in an interconnect are slight due to the relatively high thermal conductivity of interconnect materials. As mentioned earlier, the temperature of the wire segment is primarily influenced by the power dissipation of transistors and other active components on the chip and the ambient temperature. Improved heat dissipation within the chip or in the metallization layers can lower the temperature by a few degrees Kelvin only. Larger temperature changes can only be achieved with a lot of engineering effort (e.g., by

---

[1]Thermal vias are structurally similar to electrical vias, but serve no electrical purpose. Their primary function is to conduct heat vertically through the chip/die and convey it to the heat sink. A thermal wire is used to spread heat in the lateral direction.

means of external cooling, such as liquid cooling or thermoelectric cooling) or by reducing the power dissipation.

The preceding discussions show us that limiting the current density (option 1) and temperature (option 2) are largely unsuitable as "control variables" in layout design for preventing EM damage in future generations of integrated circuits. We will not therefore deal with these measures in detail in the remainder of the book.

The last three measures are more promising and deserve more attention. They are:

- layout modifications,
- material modifications, and
- new materials.

These novel approaches will be examined in detail in Chap. 4. The main focus will be on the first measure, i.e., using layout modifications to prevent EM damage, as such measures can often be integrated into existing design techniques and tools without undue difficulty. We will also discuss possible uses of material modifications and alternative materials, such as carbon nanotubes, and how they impact layout design. We will explain the effects of the measures and provide guidance for circuit designers to apply them.

References

[Bl69a]   J.R. Black, Electromigration—a brief survey and some recent results. IEEE Trans. Electron. Devices **16**(4), 338–347 (1969). https://doi.org/10.1109/T-ED.1969. 16754

[Bl69b]   J.R. Black, Electromigration failure modes in aluminum metallization for semiconductor devices. Proc. IEEE **57**(9), 1587–1594 (1969). https://doi.org/10. 1109/PROC.1969.7340

[Br04]   H.F. Brocke, *Finite-Elemente-Analyse von modernen Leitbahnsystemen*, Ph.D. thesis, Universität Hannover, 2004

[El16]   Elmer (2016), https://www.csc.fi/web/elmer, last retrieved on 1 Jan 2018

[JEP119A]   EIA/JEDEC Publication JEP119A, A Procedure for Executing SWEAT, JEDEC Solid State Technology Association, Aug 2003 [Online]. Available: http://www. jedec.org

[JESD61]   EIA/JEDEC Standard JESD61A.01, Isothermal Electromigration Test Procedure, JEDEC Solid State Technology Association, Oct 2007 [Online]. Available: http:// www.jedec.org

[JP001]   JEDEC/FSA Joint Publication JP001.01, Foundry Process Qualification Guidelines (Wafer Fabrication Manufacturing Sites), JEDEC Solid State Technology Association, May 2004 [Online]. Available: http://www.jedec.org

[JK14]   G. Jerke, A.B. Kahng, Mission profile aware IC design—a case study, in *Proceedings of the Design, Automation and Test in Europe Conference and Exhibition (DATE)*, article no. 64 (2014). https://doi.org/10.7873/date.2014.077

[JL04]   G. Jerke, J. Lienig, Hierarchical current-density verification in arbitrarily shaped metallization patterns of analog circuits. IEEE Trans. CAD Integr. Circ. Syst. **23**(1), 80–90 (2004). https://doi.org/10.1109/tcad.2003.819899

[JLS04]   G. Jerke, J. Lienig, J. Scheible, Reliability-driven layout decompaction for electromigration failure avoidance in complex mixed-signal IC designs, in *Proceedings of the 41st Annual Design Automation Conference* (2004), pp. 181–184. https://doi.org/10.1145/996566.996618

[JL10]   G. Jerke, J. Lienig, Early-stage determination of current-density criticality in interconnects, in *Proceedings of the 11th IEEE International Symposium on Quality Electronic Design (ISQED)* (2010), pp. 667–774. https://doi.org/10.1109/isqed.2010.5450505

[KLMH11]   A.B. Kahng, J. Lienig, I.L. Markov, et al., *VLSI Physical Design: From Graph Partitioning to Timing Closure*, Springer, ISBN 978-90-481-9590-9, (2011). https://doi.org/10.1007/978-90-481-9591-6

[KMJ13]   A. Krinke, M. Mittag, G. Jerke, et al. Extended constraint management for analog and mixed-signal IC design, in *IEEE Proceedings of the 21th European Conference on Circuit Theory and Design (ECCTD)* (2013), pp. 1–4. https://doi.org/10.1109/ecctd.2013.6662319

[Li89]   B.K. Liew, N.W. Cheung, C. Hu, Electromigration interconnect lifetime under AC and pulse DC stress, in *Proceedings of the 27th International Reliability Physics Symposium (IRPS)* (1989), pp. 215–219. https://doi.org/10.1109/relphy.1989.36348

[LJ05]   J. Lienig, G. Jerke, Electromigration-aware physical design of integrated circuits, in *Proceedings of the 18th International Conference on VLSI Design* (2005), pp. 77–82. https://doi.org/10.1109/icvd.2005.88

[Lie06]   J. Lienig, Introduction to electromigration-aware physical design, in *Proceedings of the 2006 International Symposium on Physical Design (ISPD)* (ACM, 2006), pp. 39–46. https://doi.org/10.1145/1123008.1123017

[Lie13]   J. Lienig, Electromigration and its impact on physical design in future technologies, in *Proceedings of the 2013 ACM International Symposium on Physical Design (ISPD)* (ACM, 2013), pp. 33–40. https://doi.org/10.1145/2451916.2451925

[LJ03]   J. Lienig, G. Jerke, Current-driven wire planning for electromigration avoidance in analog circuits, in *Proceedings of the 2003 Asia and South Pacific Design Automation Conference (ASP-DAC)* (2003), pp. 783–788. https://doi.org/10.1109/aspdac.2003.1195125

[LKL+12]   C.-H. Liu, S.-Y. Kuo, D.T. Lee, et al., Obstacle-avoiding rectilinear Steiner tree construction: a Steiner-Point-Based algorithm. IEEE Trans. Comput. Aided Des. Integr. Circuits Syst. **31**(7), 1050–1060 (2012). https://doi.org/10.1109/tcad.2012.2185050

[LM06]   J. Li, H. Miyashita, Post-placement thermal via planning for 3D integrated circuit, in *IEEE Asia Pacific Conference on Circuits and Systems (APCCAS)* (2006), pp. 808–811. https://doi.org/10.1109/apccas.2006.342144

[Ma89]   J.A. Maiz, Characterization of electromigration under bidirectional (BC) and pulsed unidirectional (PDC) currents, in *Proceedings of the 27th International Reliability Physics Symposium (IRPS)* (1989), pp. 220–228. https://doi.org/10.1109/relphy.1989.36349

[NL09]   A. Nassaj, J. Lienig, G. Jerke, A new methodology for constraint-driven layout design of analog circuits, in *Proceedings of the 16th IEEE International Conference on Electronics, Circuits and Systems (ICECS)* (2009), pp. 996–999. https://doi.org/10.1109/icecs.2009.5410838

[Pi94]   D.G. Pierce, E.S. Snyder, S.E. Swanson, et al., Wafer-level pulsed-DC electromigration response at very high frequencies, in *Proceedings of the International Reliability Physics Symposium (RELPHY)* (1994), pp. 198–206. https://doi.org/10.1109/relphy.1994.307836

[SL15]   J. Scheible, J. Lienig, Automation of analog IC layout—challenges and solutions, in *Proceedings of the International Symposium on Physical Design (ISPD)* (ACM, 2015), pp. 33–40. https://doi.org/10.1145/2717764.2717781

[WCC+05]   W. Wessner, H. Ceric, J. Cervenka, et al., Dynamic mesh adaptation for three-dimensional electromigration simulation, in *International Conference on Simulation of Semiconductor Processes and Devices (SISPAD)* (2005), pp. 147–150. https://doi.org/10.1109/sispad.2005.201494

[WDY03]   K. Weide-Zaage, D. Dalleau, X. Yu, Static and dynamic analysis of failure locations and void formation in interconnects due to various migration mechanisms. Mater. Sci. Semicond. Process. **6**(1–3), 85–92 (2003). https://doi.org/10.1016/S1369-8001(03)00075-1

Chapter 4
Mitigating Electromigration in Physical Design

The previous Chap. 3 introduced electromigration-aware design flows, for example, by identifying areas in today's design methodologies where inserting "electromigration awareness" is required. This chapter, Chap. 4, describes in detail the EM-inhibiting effects upon which this "electromigration awareness" is based and introduces further effects and related measures. We also consider material-related options to reduce EM, like surface passivation, and the use of EM-robust materials, such as carbon nanotubes (CNTs).

The goal of this chapter is to summarize the state of the art in EM-mitigating effects. We do this by providing "measures" that expand the approved current-density limits, so they are not exceeded. This knowledge can be applied by a circuit designer to increase current-density limits locally.

We determine parameters for every measure, which enable them to be used easily; we also provide detailed advice for applying each method. We will show how approved current densities can thus be increased at critical points, by means of local layout modifications. Our intent is to facilitate an EM-robust layout design in EM-critical technology nodes, such as represented by the yellow area in Fig. 1.6 (Chap. 1).

4.1 Overview of Presented Measures and Effects

In Chap. 3, we introduced components of an electromigration-aware design flow. With this knowledge as a foundation, we now present in this chapter further measures that allow the circuit designer to consider and mitigate potential electromigration threats. These measures are organized into the following topics:

- layout modifications,
- material modifications, and
- new materials.

© Springer International Publishing AG 2018
J. Lienig and M. Thiele, *Fundamentals of Electromigration-Aware Integrated Circuit Design*, https://doi.org/10.1007/978-3-319-73558-0_4

Electromigration-prevention measures are based on different physical phenomena, called EM-inhibiting effects, or simply *effects*, below.

First, we will discuss *layout modifications*. The most prominent parameters here are the interconnect width and the via size, as both directly impact current density. The crystal lattice structure of the interconnect (Sect. 2.4.1, Chap. 2) also depends on the width. The bamboo effect in eponymous crystal lattice structures is a particularly useful tool to counter the effects of electromigration. This effect will be analyzed in Sect. 4.2.

In addition to the width, the length of interconnects is easiest to modify in the layout. In terms of EM, non-trivial critical length effects are observed, arising from mechanical stresses in the interconnect and the resulting stress migration (Sect. 2.4.4, Chap. 2). Section 4.3 deals with such length effects, including also the Blech effect.

The actual "allowed" lengths for utilizing critical length effects depend on various constraints. One of these constraints can be explained by the difference between via-above and via-below configurations (Sect. 4.4). As we will see, the admissible length depends in part on the direction from which a segment is contacted by vias.

Reservoirs are another way of modifying allowed lengths (Sect. 4.5). The length or shape of an interconnect can be altered to create redundancies that increase the time to failure, or improve the reliability of a previously unreliable configuration.

The use of double/multiple vias also serves to increase reliability (Sect. 4.6). They boost the yield by reducing current density and providing redundancy. The reservoir effect contributes significantly to EM robustness here as well.

Changes in current frequency, as specified by technology advances, result in EM-behavioral changes (Sect. 4.7). There are two effects due to current frequency: (i) self-healing and (ii) the skin effect, which impacts the local current density (Sect. 2.4.3, Chap. 2).

All of these effects and measures differ in their respective impacts depending on the technologies and materials used. The bamboo effect, especially, depends on the relationship between the activation energies and the different diffusion mechanisms. There are marked differences in this regard between the metals used in the interconnects. The barrier materials must be taken into account, as well. Further, the critical length effects depend on the mechanical properties of the dielectric. We therefore discuss all important materials in the routing layers in Sect. 4.8, thus pursuing the *material-modification* approach.

In order to push the boundaries of EM robustness further out beyond classical technologies, a third electromigration-prevention measure is presented in this chapter (Sect. 4.9), termed *new materials*. As we will see, high current densities are achievable without EM damage with the advent of new materials, especially CNTs.

The design parameters that have a bearing on the effects examined in the following sections are shown in Table 4.1. We can see from the table that almost all effects are material- and technology-dependent.

Table 4.1 Relationships between design parameters (top line) and different EM-influencing effects and measures (left column)

	Length L	Width W	Frequency f	Material	Technology
Bamboo effect, Sect. 4.2		✓		✓	✓
Blech effect, Sect. 4.3	✓			✓	✓
Via effects, Sect. 4.4	✓			✓	✓
Reservoir effect, Sect. 4.5	✓	✓	✓	✓	✓
Via configuration, Sect. 4.6	✓	✓			✓
Self-healing, Sect. 4.7			✓		
Passivation, Sect. 4.8				✓	✓
Immunity, Sect. 4.9				✓	✓

4.2 Bamboo Effect

4.2.1 Fundamentals

The bamboo effect is based on the elimination of grain boundaries in the interconnect that could act as diffusion paths for EM. The same favorable properties with regard to EM that are present in a monocrystal (no grain boundaries parallel with the direction of diffusion, Sect. 2.4.1) are achieved far more easily with bamboo structures, since, in contrast to the monocrystal, many more crystal nuclei are permitted. A coarse-grained crystal lattice structure is grown by a process known as *tempering*, that is, holding the interconnects at an elevated temperature over a longer period, followed by slow cooling (thermal annealing [Hoa88]) after metal deposition.

Interconnect dimensions—mainly the cross-section, i.e., wire height and width—are a critical factor in the formation of a given crystal lattice structure within the interconnect. If the wire width is reduced while keeping the current and temperature constant, the time to failure changes as shown in Fig. 4.1.

Tracing the curve from right to left, we observe first a reduction in the wire's reliability, i.e., its time to failure, which is caused by the corresponding increasing current density. This is caused by the formation of near-bamboo crystal lattice structures, where divergences in the diffusion flow occur as a result of blocking grains and triple points (Sect. 2.4.1). Second, in the left part of the curve we see an increase in the time to failure below a width limit that is about half the grain size [KCT97], as there is less diffusion overall due to the bamboo structure. In the bamboo structure, there are no more continuous diffusion paths in the form of grain boundaries, as the boundaries are only perpendicular to the direction of diffusion. This leads to less diffusion and thus to a lower failure probability, as illustrated by the favorable rise of the curve in the left-hand part of Fig. 4.1.

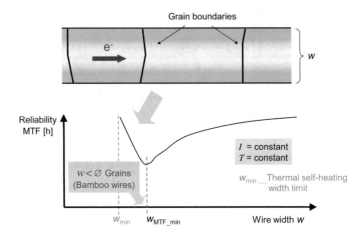

Fig. 4.1 Reducing wire width to less than the average grain size improves interconnect reliability with regard to electromigration. So-called bamboo wires are characterized by grain boundaries which are perpendicular to the direction of the electron wind and thus permit only limited grain-boundary diffusion

There is however a limit to the scale of this effect. The growing current density and the disproportionate increase in resistance of the interconnect caused by the reduction in width result in thermal damage (Joule heating). The increase in resistance is not a linear function of the reciprocal cross-section, but it increases at a higher rate under the growing influence of edge and scattering effects [ITR14, ITR16, Aro03]. Furthermore, the minimum wire (track) width is always defined by the technology constraints.

Interconnects carrying high currents (preferably power supply nets) can benefit in another way from the bamboo effect. The interconnect can be divided into a number of parallel bamboo structures, with parts of the current flowing through the different structures. This *wire slotting* or *cheesing* process (Fig. 4.2) is primarily used to comply with the constraints of chemical-mechanical polishing (CMP). There are CMP guidelines that stipulate maximum copper widths and the surface ratio between metal and dielectric to ensure uniform material abrasion. The reduction in width for high ampacity interconnects facilitates the formation of bamboo structures, whose beneficial side effect reduces electromigration.

Fig. 4.2 Differently slotted wires, left rectangular slots, right octagonal cutouts, that support the bamboo effect [KRS+97, KWA+13]

Additional process steps are required to successfully produce bamboo structures. For example, crystal growth and lattice rearrangement triggered by thermal annealing [Hoa88] lead to larger grains in the lattice, which increases the probability of producing the desired bamboo structures.

One caveat is that the bamboo effect is highly material-dependent. The effects are very extensive in aluminum, as the grain-boundary diffusion is the predominant type of diffusion in the overall diffusion flow. Bamboo structures have very little impact on the time to failure of copper, as surface diffusion is the determining process here (see Table 2.1, Chap. 2).

There are reliability risks associated with these structures too, if the creation of a bamboo structure is not assured. Such a lattice would contain triple points or individual blocking grains (see Fig. 2.10, Chap. 2). Large divergences in the diffusion flow will result causing increased and concentrated occurrences of voids and agglomerations. It is therefore essential that complete bamboo structures are created in the material treatment process, in order for this process to be effective.

4.2.2 Applications

The bamboo effect is particularly suited for aluminum wiring with track widths less than 2 μm, as was shown with an Al-0.5%Cu alloy, among others [VS81]. Bamboo structures can be created in copper with track widths of less than 1 μm by thermal annealing (3 h at 400 °C) [HRL99]; however, they have a positive effect only if the surface diffusion is largely eliminated. Practical solutions in this regard are presented in Sect. 4.8.

Bamboo-type structures are effective only with grain sizes greater than twice the track width ($w/D_{50} < 0.5$) [KCT97]. There are several studies in the literature describing processes to create large grains; for example, copper bamboo structures have been created in 0.5-μm-wide interconnects by scanned laser annealing (SLA) with grain sizes up to 4 μm [HRT01]. In contrast, regular annealing generated mean grain sizes of only 0.13 μm at 275 °C for 24 h. In addition, monocrystalline and bamboo structures were produced in aluminum by annealing with the help of sodium chloride (NaCl) layers [JT97], whereby interconnect widths up to 2 μm were allowed.

Unfortunately, the various annealing processes used in the tests noted above are unsuitable for manufacturing, as the temperatures required are too high, or the technologies required are too elaborate and expensive. In particular, annealing at 575 °C, as in [JT97], is unsuited for semiconductor transistors; NaCl cannot be used as a crystal nucleus [JT97], nor can SLA [HRT01] be used, due to its incompatibility with wafer fabrication.

When we consider overall trends in the miniaturization of integrated circuits, it is clear that the ever-smaller structure sizes facilitate bamboo structures in interconnects. This applies especially to the *Damascene* process, as it allows the

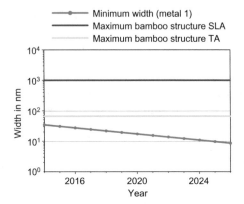

Fig. 4.3 Comparison of actual occurring track widths (metal 1, *blue line*) with boundary values for bamboo structures (*red, yellow*), SLA = scanned laser annealing [HRT01], TA = thermal annealing at 100 °C [HOG+12], minimum width = half pitch [ITR14]. As illustrated, minimally dimensioned track widths in bamboo structures can be produced with suitable processes in the bottom metal layers with current technologies

recrystallization of large contiguous metal layers during annealing before CMP. This increases the probability of obtaining bamboo structures in the interconnects.

Figure 4.3 shows that both current structural widths and upcoming technologies allow bamboo structures to be produced in minimally dimensioned interconnects in the bottom metal layers. The necessary technology steps (annealing) must be integrated in the manufacturing process to achieve this. In the case of copper, however, EM properties only benefit from bamboo structures if surface diffusion is disabled (Sect. 4.8.3).

4.3   Critical Length Effects

4.3.1   Fundamentals

According to the Blech effect, named after Ilan Asriel Blech [Ble76], the time to failure of an interconnect in an electromigration test depends on the length of the interconnect. In this context, no visible damage in the form of voids has been observed in interconnects shorter than a critical length that depends on the current density. The reason for this phenomenon is the negative feedback between electromigration and stress migration, described in Sect. 2.5 (Chap. 2). The material transport caused by electromigration ensures that the void concentration at the cathode end increases, while it decreases at the anode. Stress migration counteracts this process (Fig. 4.4), effectively neutralizing it if the interconnect length is shorter than a critical length (whose value depends on the current density).

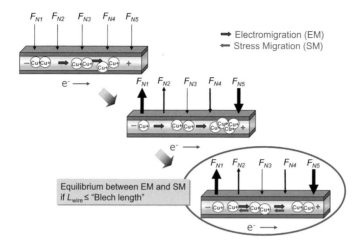

Fig. 4.4 Simplified illustration of the interaction between electromigration and stress migration. The latter is caused by the mechanical stress buildup in a short wire. This reversed migration process essentially neutralizes the material flow due to electromigration

The discussion in this section refers to individual interconnect segments. A *segment* here is a part of a conductor that is terminated either by two interconnects in another layer (vias or contacts), or by branches within the plane of the segment, and that contains no other branching in-between (Fig. 4.5). A wire segment is thus equivalent to the edge of a graph in graph theory, where the nodes in the graph are represented by connections, branches, or vias. When analyzing DC currents, each segment contains an anode end and a cathode end. As is common, the cathode-end potential is negative and the anode-end potential is positive.

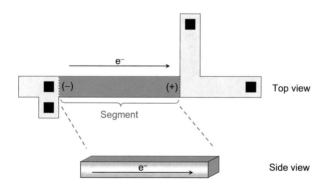

Fig. 4.5 Illustration of a wire (interconnect) with a segment highlighted with (−) = cathode, (+) = anode, and vias (black)

Fig. 4.6 Mechanical stress profile over time across the length of a segment with the Blech effect (time sequence: blue, yellow, orange, red). Note that the critical mechanical stress $\sigma_{critical}$, which leads to the creation of a void, is not exceeded in order to fulfill the equilibrium condition as specified by the Blech effect

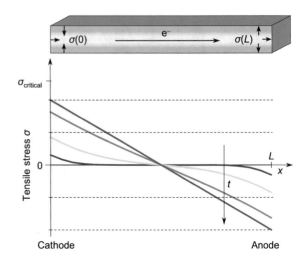

The electromigration-induced material transport produces a mechanical stress gradient within the segment (Fig. 4.6), which is the driving force behind stress migration. In this case, the gradient counteracts the material flow produced by electromigration. Hence, the resulting diffusion flow continuously decreases. If a steady state is established, the result is a linear stress characteristic with a constant gradient along the length of the segment (see Fig. 4.6, red curve) [WHM+08]. This stress characteristic means that the diffusion flows from electromigration and stress migration negate each other across the entire segment length. If the critical mechanical stress $\sigma_{critical}$, which leads to the creation of a void, is not exceeded, the equilibrium condition as specified by the Blech effect is satisfied. Permanent damage to the segment due to electromigration is not expected in this case.

Blech discovered that this equilibrium depends on the current density j and the length L of the segment. For every interconnect cross-section, there is a critical product $(jL)_{Blech}$ below whose value no voids are formed. This critical product is determined by several factors, which include among others, the fabrication technology. Setting the condition $J_a = 0$ in the equation for the one-dimensional diffusion flow [Eq. (2.18) in Chap. 2], we get the expression for the Blech effect as follows:

$$(jL)_{Blech} = \frac{\Omega \cdot \Delta\sigma}{e \cdot z^* \cdot \varrho}. \tag{4.1}$$

The parameter $(jL)_{Blech}$ is the maximum permissible product ("length") at which the Blech effect occurs, and Ω is the atomic volume. The difference $\Delta\sigma$ is the difference between the respective mechanical stresses at the anode and cathode. Further, e is the elementary charge, z^* the effective charge of copper, and ϱ the specific electrical resistance of the interconnect.

If this critical length or the critical product jL is exceeded, void growth is initiated. If minor damage in the form of voids is tolerated, the permissible current density can be increased. Voids arise, as a rule, before extrusions start to form at the anode, because (i) the interconnect is under tensile stress in the initial state and (ii) the critical threshold for tensile stresses is lower than for compressive stresses in terms of absolute values. The opposite side is thus more resistant to damage even though voids have already been created [WGT+08].

Under certain conditions, these interconnects can still have an almost limitless time to failure even if the critical product jL is exceeded. This is due to another effect, the so-called *void-growth saturation*, which is also caused by the negative feedback of stress migration. In this case, the mechanical stress at the cathode end exceeds the critical value σ_{critical}, at which voids are created (Fig. 4.7). When voids are formed, the mechanical tensile stress at the cathode disappears; this causes a buildup of compressive stress at the anode end, which leads to a stress gradient across the length of the interconnect segment as long as the critical value for forming extrusions has not been reached. The size of the void that is formed determines whether the interconnect fails or not.

Different void volumes cause a critical increase in the interconnect resistance, depending on interconnect geometry and the location where the voids are created

Fig. 4.7 Mechanical stress profile over time across the length of a segment at saturated void growth [WHM+08]. Note the time sequence blue, yellow (void creation), orange, red, dark red. Voids are created if the mechanical stress at the cathode end exceeds the critical value σ_{critical}. In this case, the mechanical tensile stress at the cathode disappears (orange), leading to a buildup of compressive stress at the anode, with a stress gradient across the length of the segment (red, dark red)

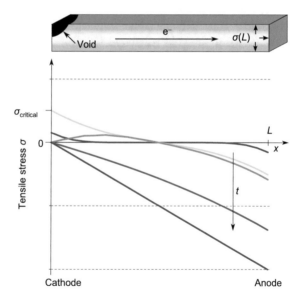

(see also Sect. 4.4). Similar to the Blech effect, the critical length or the critical current can be expressed by means of the product $(jL)_{sat}$ as follows [Tho08]:

$$(jL)_{sat} < \frac{\varrho/A}{\varrho_1/A_1} \cdot \frac{\Delta R_{fail}}{R} \cdot \frac{2\Omega B}{ez^*\varrho}. \tag{4.2}$$

In addition to the variables from Eq. (4.1), the terms ϱ/A and ϱ_1/A_1 are the relationship between resistance and cross-sectional area of the interconnect and the barrier, respectively (the subscript 1 stands for liner), ΔR_{fail} is the maximum permissible resistance change, R the output resistance, and B the effective Young's modulus for the interconnect environment, which depends proportionately on the properties of the dielectric and the barrier material.

The parameter jL symbolizes the link between electromigration and stress migration. In the equilibrium case, a gradient of mechanical stress, which is dependent on the current density, is required to trigger a compensating stress migration. The product jL indicates whether the critical stress has been reached or not, assuming a constant gradient and an actual critical mechanical stress that leads to void formation.

The product jL is not considered a useful metric in all the research literature, because results based upon it are too conservative. Lamontagne et al. [LDP+09] suggest replacing the criterion with jL^2, which would lead to less restrictive boundaries especially for short segments of less than 20 µm. This criterion is proportional to the void volume as follows:

$$V_{sat} = \frac{e \cdot z^* \cdot \varrho \cdot A_{Cu}}{2 \cdot \Omega \cdot B} \cdot jL^2. \tag{4.3}$$

Besides the variables in Eq. (4.2), A_{Cu} is the cross-sectional area of the interconnect.

According to the authors of [LDP+09], this criterion better describes the failure mode, as it is proportional to the absolute resistance change ΔR_{sat}. Due its associated voltage drop, the absolute resistance change is a better measure of the failure than the relative resistance change. Transistors may not function properly if a specific voltage drop (IR drop) is exceeded. This proportional relationship to the absolute resistance change makes it easier to differentiate between critical and non-critical current densities.

The critical volume of a void is the threshold above which the interconnect is permanently damaged. This volume corresponds to a critical resistance rise that either (i) causes a voltage drop across the interconnect to impair proper functioning of the circuit or (ii) leads to severe thermal damage (Joule heating) due to the power dissipation in the interconnect. Electromigration acceleration due to a positive feedback loop, described in Chap. 2 (see Fig. 2.4), is another aspect that must not be ignored. Here, the current density increases in the damaged zone thereby exacerbating the damage. The critical resistance change can be "translated" to a

Fig. 4.8 Model of an
interconnect with isotropic
void-volume growth (side
view). The entire
cross-section is depleted
when the void length equals
the interconnect height/width
[LDP+09]

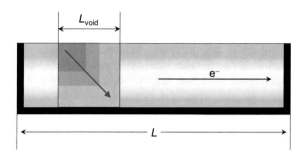

critical void volume depending on the exact location of the site where the voids are created.

Damage by an inadmissibly high resistance change typically occurs only if the copper cross-section is completely broken. All the current then flows through the barrier, especially the metal liner, surrounding the interconnect. There is a measurable resistance change, due to the low residual cross-section coupled with a higher specific resistance. If we assume, when modeling this phenomenon, that void growth is approximately isotropic (Fig. 4.8), then the volume $V_{void} = H^2 \cdot W$ is the volume at which the void spreads to the full extent of the copper cross-section of the interconnect, where H is the height and W the width of the wire. An increase in resistance in realistic wire lengths can only be identified when this critical value is exceeded, as the barrier material in the environment of the voids then becomes the sole conduit of the current. The barrier has a very low cross-sectional area and a relatively high specific resistance.

As some copper is available in the presence of a reduced void volume, a very low interconnect resistance still exists. In this scenario, where the void volume is less than $H^2 \cdot W$ for saturated growth, the wire is EM robust. Such wires are said to be immortal.

As the precise geometry of voids is never known and their analysis is based on statistics, a safety factor must be included in the calculations. A simple model for

Fig. 4.9 Model of an
interconnect with
void-volume growth after
the entire interconnect
cross-section has been
depleted (side view)
[LDP+09]

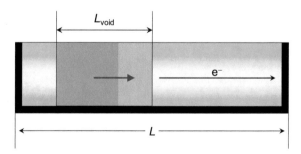

the relation between void volume and resistance change is defined in [LDP+09]. This model assumes that the void has spread across the entire cross-section of the interconnect and that only the remaining outer layer made of barrier material contributes to the conductivity (Fig. 4.9). The relation between void volume V_{sat} and resistance change ΔR_{sat} is given by:

$$\Delta R_{\text{sat}} = \frac{V_{\text{sat}} \cdot \varrho_b}{A_{\text{Cu}} \cdot A_b}. \tag{4.4}$$

The quantity ϱ_b is the specific electrical resistance of the metal liner, A_{Cu} is the cross-sectional area of the interconnect, and A_b is the cross-sectional area of the barrier. Hence, the void volume V_{sat} is calculated from the resistance change ΔR_{sat} as follows:

$$V_{\text{sat}} = \frac{A_{\text{Cu}} \cdot A_b \cdot \Delta R_{\text{sat}}}{\varrho_b}. \tag{4.5}$$

The void volume is also a measure of the quantity of material that may be transported by diffusion before critical damage occurs. This quantity of material is a function of the buildup of mechanical compressive stress at the anode. Should the compressive stress in relation to the segment length at the critical void volume initiate sufficient stress migration, the segment is stable for the given current density.

The work of Blech [Ble76] sparked a series of practical investigations into the eponymous effect. It has emerged from these myriad studies that the effect is highly influenced by the materials and technologies used. (In contrast, Blech examined metal structures of different lengths and concluded that the median time to failure is a function of the length—thus implying a constant parameter jL.)

The effect has been investigated by numerous researchers using so-called *Blech structures*, consisting of different-length segments connected in series (Fig. 4.10).

The test segments are placed on a strip of material with low electrical conductivity, such that most of the current flows through the segments. The purpose of the test is to establish the segment length at which damage occurs depending on the current. The test can have three different outcomes: (i) the structure is destroyed, (ii) slight damage occurs, or (iii) there is no measurable structural change. Such tests yielded values between 420 and 3800 A/cm for critical Blech products jL for aluminum [Sch85];

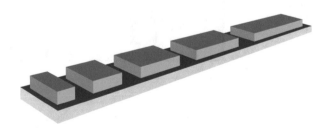

Fig. 4.10 Schematic representation of a test structure to verify the Blech length consisting of different segment lengths, connected in series on a strip of material with low electrical conductivity

values of 375–3700 A/cm were found for copper [Tho08]. Depending on the material combinations and geometrical configurations, this results in lengths of approximately 5–100 μm for copper for typical current densities.

The theory for critical length effects is examined fully in [CS11, Tho08, PAT99]. Please refer to these publications for further details and formulae.

4.3.2 Applications

The critical length effect is being deployed in part in IC design today: special rules for maximum currents are applied for segments with lengths that are definitely shorter than the Blech length. In general, these are simplified rules that are adapted for a given technology, as described below.

The following rules are defined in [Set09]:

$$I_{max}(5 \leq L \leq 10) \sim \frac{W}{L} \cdot S, \qquad (4.6)$$

$$I_{max}(L < 5) \sim \frac{W}{5} \cdot S, \qquad (4.7)$$

where W is the segment width, L the segment length, and S a user-defined de-rating factor. Equation (4.6) describes a length-dependent current-density rule for considering the Blech effect, and Eq. (4.7) defines a maximum current (density), as a margin of safety, for very short segments.

Other rules are available for the current-carrying capacity of longer segments [Set09] as follows:

$$I_{max}(2 < W < 20) \sim W \cdot \sqrt{W} \cdot S, \qquad (4.8)$$

$$I_{max}(W \geq 20) \sim W \cdot S. \qquad (4.9)$$

The permissible current (density) depends only on the interconnect cross-section in Eqs. (4.8) and (4.9). The impact of the lower time to failure in the case of small interconnect cross-sections is additionally included in Eq. (4.8). Equation (4.9) is suitable for wider interconnects with a linear current-density relationship to the width.

Obviously, the segment length should be limited in the layout, especially for high currents. Doing so will increase the number of electromigration-robust segments in the IC, based on critical length effects. All other interconnects must consist of durable segments, to ensure an ample margin of safety for the required IC time to failure.

The analyses in this section are only valid for simple segments, such as linear parts of interconnects having two terminals; supplementary approaches and methodologies are required for more complex structures. An example of a supplementary approach is the reservoir effect, which we will examine in the following sections. Both immortal segments and durable segments can be designed with critical length effects, as the transition from "immortal" to "non-durable" is gradual.

The maximum lengths for the utilization of the effect are on the order of a few microns for common current densities. These lengths should be compared with segment lengths found in real electronic circuits in order to correctly assess the applicability of the critical length effect.

The actual segment lengths and their statistical distribution can be determined in several ways. First, segment lengths can be analyzed statistically, and a mathematical model derived. Relevant models for the net length based on a model by Davis [DDM96] are presented in the literature [HC08, SNS+07, ZDM+00], and Rent's rule [LR71] is often used to calculate the number of pins required for an integrated circuit. A typical length distribution according to [SNS+07] is presented in Fig. 4.11. In this approach, the overall wire length per net (net length) must be statistically spread across the individual routing layers in order to determine a realistic segment length.

Alternatively, as a second approach, the available routing resources can also be used in the calculations. The entire routing lengths of the bottom six metal layers per chip surface area are quoted in the ITRS [ITR14]. If one spreads these across the routing layers and compares them with the number transistors per chip surface area (from the ITRS as well), one gets also an estimate of the mean segment length.

Figure 4.12 is based on the second approach, where resources are uniformly distributed among the six routing layers, and two nets per transistor are assumed. The plot represents the mean value for layers *metal 1* to *metal 6*. Mean segment lengths occurring in typical integrated electronic circuits with state-of-the-art technologies are plotted compared to the (theoretical) Blech length. Considering the model error and the statistical variance of the input data, we can assume quite a wide statistical length distribution due to the greatly simplified calculations.

The graph in Fig. 4.12 shows that the maximum lengths for utilizing critical length effects are not exceeded in many cases. The length distribution of the interconnects in Fig. 4.11 indicates that larger segment lengths make up only a

Fig. 4.11 Statistical distribution of the signal net lengths for a digital 65-nm technology with eight routing layers [SNS+07]. Resources used by power supply nets are also taken into account

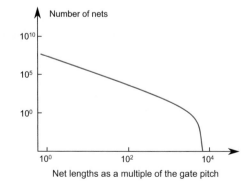

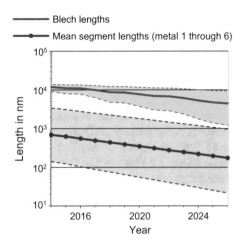

Fig. 4.12 Average segment lengths (blue) compared to the Blech length (red) for typical current densities, data from [ITR14]. Estimated variances (standard deviations) for the quantities are shown. The plots indicate that the mean lengths including standard deviations observed on ICs typically comply with the (theoretical) Blech criterion for immortal interconnects. Blech lengths are however being exceeded by an increasing proportion (up to approximately 5% by the year 2026) of the chip's routing. There is a tendency for the curves to converge even further under the influence of technological constraints

small portion of the routing structures. There are nonetheless many above-average net lengths on the chip, as well. This is due, in part, to the abundance of local routing connections within subcircuits in the lower metal layers.

The above-mentioned lengths typically depend on the logic-gate pitch. However, in global/external routing connections, where segments are longer, the lengths are related to the block and chip sizes. These connections are typically located in the upper routing layers where cross-sections are larger. The current density is lower in these cases, and segment lengths can thus be larger, based on the specified jL product.

Metal layers with small cross-sections are the subject of our investigations in this section. These are typically the first four to six layers with the smallest structures located near the substrate. The most difficult problems with electromigration occur in these layers, while very few EM problems are expected in the near future in the higher, and thus more coarsely structured, layers.

The lower first or second metal layers in digital circuits are typically dedicated to connections within the cells; as a result, only small lengths are expected in these layers. However, longer lengths that exceed the Blech criterion may exist in the third to sixth layers.

The critical length effects as described here apply to individual interconnect segments. How the effect influences the time to failure, however, is determined to a large extent by the exact mechanical boundary conditions that act on the segment. These boundary conditions are not only constrained by the segment itself and how it is embedded in the dielectric, but also by the linking to neighboring segments in the same metal layer.

In practice, the Blech effect is applied by means of geometrical rules in the layout design. Specifically, care should be taken during the routing stage that the critical product jL or jL^2, as the case may be, is not exceeded.

Utilizing the Blech effect, commercial and academic EM-aware routers are increasingly restricting the permissible length of routing segments (cf. Eq. (4.6)). Additionally, longer routing paths (net lengths) can be implemented with shorter segments by introducing layer changes. Although this measure is rarely used in current designs, it can be a useful option for very long connections. However, the trade-off between the reliability benefit from the critical length effect and the loss of reliability caused by introducing additional vias should be weighed (Sect. 4.6).

4.3.3   Linked Segments

A net consists of a number of segments, forming a chain or tree structure depending on the net topology. Neighboring segments impact the hydrostatic stress in a manner similar to reservoirs (Sect. 4.5). In addition, they also emit an atomic diffusion flow as they carry current. The sum of the effects and their impact on neighboring segments yields the total critical length effect. In other words, the entire net must be analyzed, or, when considering a segment, its neighboring segments must be analyzed as well.

We cannot therefore draw conclusions from the behavior of linear (individual) segments for the interaction between linked segments of a routed net in a metal layer. In fact, the opposite conclusion might be true. Studies, such as those of Chang et al. [CCT+06], suggest that the behavior of linked segments can be very different to that of individual segments. Hence, the current densities and the mechanical stresses at the segment nodes must be taken into account. Wei et al. [WHM+08] investigated the effects of linking numerous segments.

An attempt is made in [HGZ+06] to extend the analysis of the critical length effects across whole nets, and across the entire IC as well, with the help of hydrostatic stress. This effect is investigated by applying different currents to two connected segments in order to determine the mutual interaction of the segments w.r.t. the predicted time to failure. The following aspects should be considered in such cases:

- the length of the segment under test,
- the length of the connected segment,
- the relation of the two currents in the segments,
- the embedding of the segment in the dielectric, and
- the mechanical constraints, i.e., external stresses.

The difference between current density as the driving force for EM, and hydrostatic stress as the driving force for stress migration (SM), is that the latter also acts in the insulator that surrounds the interconnect. The EM and SM effects are balanced in the steady state, resulting in no diffusion flow. The stress must be distributed in the environment of the interconnect segment in order for this condition to be met.

Given a steady state, where there is no violation of the mechanical constraints, such as maximum tensile or compressive stress in the material or in a boundary layer, it is unlikely that the assumed current density in this segment will lead to a failure; we say "unlikely" because boundary conditions not taken into consideration in this model could facilitate a failure. Also, there is no guarantee in the steady-state model that this state can be reached solely by EM and SM.

A dynamic model of the entire process from the initial state to the steady state can ensure the robustness of a segment with somewhat more certainty. This type of model, however, is unsuited for circuit analysis, nor would it be economical, given the considerable increase in computation time required, and the high degree of uncertainty due to the impact of initial conditions and statistical deviations in the metal structures.

To sum up, the electromigration models to date are inadequate for critical length effects that are expected to be increasingly important in the future.

4.4 Via-Below and Via-Above Configurations

4.4.1 Fundamentals

As indicated in Sect. 4.3, the practical utilization of critical length effects depends on additional boundary conditions. One of these boundary conditions relates to the difference between *via-above* configurations, also called *downstream* configurations, and *via-below* configurations, also called *upstream* configurations (Fig. 4.13). Simply put, the permissible length is based on the direction from which a segment

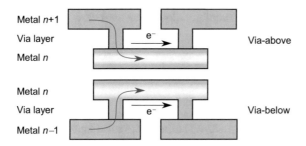

Fig. 4.13 Side views of via-above (top) and via-below configurations (bottom)

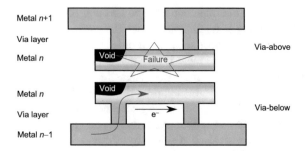

Fig. 4.14 As voids tend to occur on the top surface of the interconnect (due to the CMP process, among others), voids formed in the via-above configuration (top) lead to earlier failure than in the via-below configuration (bottom)

is contacted by vias. This effect is best understood by studying the manufacturing process for integrated circuits (see also Sect. 2.4.2, Chap. 2).

In the dual-Damascene process, the deposited copper layer, which forms the interconnect layer, is partly removed by polishing in the chemical-mechanical polishing/planarization (CMP) process [Gup09, Yoo08]. Defects in the metal surface are caused by this process which cannot be removed subsequently. The result is a high defect density and vacancy concentration on the top surface of the interconnects. If these phenomena are combined with a dielectrical encapsulation material with poor adhesion, the top surface is more exposed to electromigration damage. This is why voids tend to occur on the top surface of an interconnect. Thus, if an interconnect is contacted from above (via-above, downstream), the void often occurs directly in the contact zone between via and interconnect. The result is that a small void can cause the interconnect to fail, while the upstream configuration can tolerate more volume loss before damage occurs (Fig. 4.14).

We can quantify the effect by examining interconnects where catastrophic destruction of the interconnect is essentially prevented by void-growth saturation (Sect. 4.3). In this scenario, the current-density-length product can be increased tenfold with a via-below configuration. The via-below configuration thus fully leverages the critical length effect.

4.4.2 Parameters

The critical values for jL (Sect. 4.3) differ quite substantially between upstream and downstream configurations because of their different critical void volumes. Values quoted in [Tho08] differ by a factor of ten; for example, a value of 375 A/cm is

Fig. 4.15 Comparison of actual and expected segment lengths (blue line) with boundary values for via-below and via-above configurations [ITR14]. Despite the fact that the two critical Blech lengths are always greater than the *mean* actual segment lengths, there are potentially more nets with via-above configurations that exceed the critical lengths than there are with via-below configurations (cf. Figs. 4.12 and 5.1, Chap. 5)

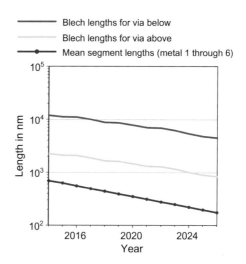

given for jL in [Tho08] for the via-above configuration, while values of up to 3700 A/cm are reached for the via-below configuration. This corresponds to lengths of 7.5 and 74 μm for via-above and via-below configurations, respectively, for typical current densities of 5×10^5 A/cm^2.

If we compare the actual mean segment lengths measured on integrated circuits with the potential (theoretical) critical lengths of via-below and via-above configurations (Fig. 4.15), we observe that the two critical lengths are always greater than the actual *mean* lengths found on the chip. However, there are considerably more nets with via-above configurations that potentially exceed the critical lengths because of the above-mentioned length difference between the two types of configurations and considering the length distribution (see Fig. 4.11). Hence, the differences between via-below and via-above configurations must be taken into account when permissible current densities and segment lengths are defined.

The materials used (Sect. 4.8) also have a significant impact on these issues. However, regardless of the material, the top surface of the metal layer continues to suffer most from EM damage with current technologies.

4.4.3   Applications

There is a definite tendency for voids to occur at the top surface of an interconnect. Besides the aforementioned CMP-induced defects, this is due to different activation energies for electromigration at the various interconnect surfaces [SMS+07, HLK09, HGR06]. The latter is caused by the different types of barrier materials and the CMP process on the top surface.

Hence, the critical segments of an interconnect should, if possible, be configured as via-below configurations in the design flow. Furthermore, different critical *jL* products should be applied respectively to via-below and via-above configurations.

Ultimately, the type of configuration impacts the critical void size. This size is the limit up to which the segment remains intact. Interconnect durability benefits if this limit is maximized for a segment by means of the geometrical layout. The actual void size must stay below the critical limit or, better still, growth saturation should be reached in order to meet the required time to failure for a given interconnect (Sect. 4.3.1).

For via-above configurations, an overlap between the metallic barriers (metal liners) at the via and the interconnect underneath is also recommended. This limits the resistance increase caused by voids under the via [MS13].

4.5 Reservoirs

4.5.1 Fundamentals

Reservoirs are interconnect segments that are connected with other segments and are typically not current carriers themselves. They nonetheless modify the effect of EM on current-carrying segments.

The basis of the reservoir effect is the critical length effect and the effect of mechanical stresses described in Sect. 4.3. Specifically, the shifting of the prevailing equilibrium during void-growth saturation is a key enabler of the reservoir effect.

Reservoirs provide material for diffusion, thus preventing void growth from damaging the interconnect. Larger void volumes can be permitted, while sufficiently high mechanical stresses occur to saturate the void growth by stress migration. As a result, the lengths and/or current densities permitted by the critical length effects are thus increased.

Besides evaluating individual segments, entire nets comprising many connected segments can also be considered when analyzing reservoirs. Reservoirs can be treated as short, non-current-carrying interconnect segments [WHM+08]. The effect reservoirs have on the connected segments thus exist as well between neighboring segments in a routed net.

4.5.2 Sources and Sinks

Source- and sink-type reservoirs differ in their effects (Fig. 4.16). A source-type reservoir makes material available and enables the development of voids without interrupting the current flow. It also increases the critical void volume at which a

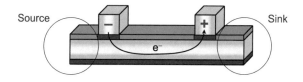

Fig. 4.16 Source- and sink-type reservoirs of a via-above segment

measurable change in conductor properties is observable. A source thus increases the time to failure of a segment, and, depending on the current, can even render the segment immortal.

In contrast, a sink "makes space available" in the form of vacancies for the transportation of interconnect material. On the one hand, this reduces the probability that the critical mechanical stress threshold for the development of hillocks and whiskers at the anode is exceeded. However, higher absolute values of mechanical stress are usually needed for the creation of hillocks and whiskers than with voids. Thus, this positive circumstance is not fully effective. On the other hand, the accumulated mechanical stress for the equivalent transported quantity of materials is reduced, because the material is distributed over a larger volume at the end of the interconnect. As a consequence, the quantity of materials and hence the void volume at the cathode—where stress migration and electromigration are balanced—increases. A sink therefore increases the probability of a failure.

As discussed in Sect. 4.3, the stress migration caused by the mechanical stress gradients ensures a stable interconnect status without further void growth in a manner similar to the critical length effects. This equilibrium can still establish itself due to the reservoir with larger void volumes, without critical damage to the interconnect.

4.5.3 Reservoir Types

There are a variety of reservoir types. The most basic type is the *end-of-line reservoir* (Fig. 4.17). It has the smallest footprint and the least difference to a layout without reservoirs. It consists of an enlarged overlap at the via (see Fig. 4.16) and has the advantage that the layout is based on the predefined grid and remains compatible with techniques such as double or triple patterning. (Double or triple patterning increases the resolution by spreading the layout structures across two or three separate masks.)

Side reservoirs, on the other hand, do not always meet double-patterning guidelines. Side reservoirs give rise to two-dimensional structures, that is, branches are introduced, as the reservoirs are not aligned in the segment direction (Fig. 4.18).

The advantage of these latter structures is that voids, occurring inside a segment, are "intercepted" by side reservoirs without jeopardizing the connection to the via [MGL+11]. In contrast, in the case of reservoirs at the end of the line, a void "passing by" would cause a failure at a via before reaching the reservoir.

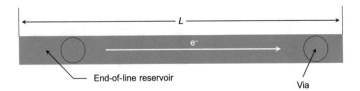

Fig. 4.17 Top view of an end-of-line reservoir of an interconnect segment

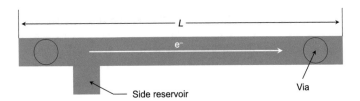

Fig. 4.18 Top view of a side reservoir

Reservoirs are also found as a by-product of multiple parallel vias; they occur in the gaps between the vias (Sect. 4.6) and produce source and sink reservoirs. Special attention should be given to this phenomenon in nets subjected to AC loads, as outlined in the next Sect. 4.5.4.

4.5.4 Applications

If reservoirs act exclusively as a source, they improve the electromigration characteristics of segments where voids could arise. The critical void volume is enlarged, as a void causes no damage if it is located within the reservoir.

A reservoir can only exclusively act as a source if current in the segment always flows in the same direction. This is not the case if the currents in the segment are alternating currents, as a reservoir can be created as a source or sink, depending on the direction of the current. The positive source effect is inferior to the effect of the sink if low-k dielectrics, that have a lower Young's modulus, are deployed. In such scenarios, reservoirs should be avoided.

The permissible current-density-length product jL can be increased by utilizing the positive effects of reservoirs (Fig. 4.19). A fivefold average increase in the permissible length at which the Blech effect can be applied can be achieved for the same current density [HRM08].

The negative influence of the mechanical properties of low-k dielectrics can be counteracted by introducing metal structures in the proximity of the anode and cathode or by mechanically reinforcing the dielectric locally [Tho08]. Reservoirs are modeled in [NSMK01] in order to calculate the time to failure as a function of

Fig. 4.19 Comparison of (actual and expected) segment lengths with (theoretical) boundary values with and without a reservoir (via-below configuration). The graph is based on mean segment lengths derived from the routing density and transistor density and indicates a significant increase in the permissible length at which the Blech effect can be applied. Data from [ITR14]

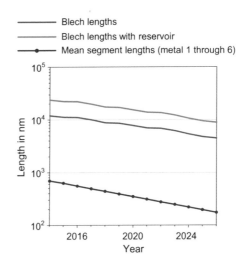

the size of the reservoirs: the time to failure for the segment scales with the reservoir surface area. In the case of the length of an end-of-line reservoir, on the other hand, an optimum length was found that correlates with a maximum time to failure [LNW10, TF12].

Time to failure, and hence wire reliability, does not benefit from sinks, as there is a smaller buildup of mechanical stress gradients for the same transported quantity of materials. Hence, larger voids are produced in the balanced state. In addition, there is a higher probability of failure with low-*k* dielectrics, as the mechanical stresses are further reduced in this case. The impact of an existing source is thus over-compensated, and the sink also reduces the time to failure under AC loads [Tho08].

A reservoir can exert different positive and negative effects in a segment carrying alternating currents, and it is important to consider which of these effects pre-dominate [HRM08]. Reservoirs exert a negative impact on the time to failure in the presence of alternating currents, especially with low-*k* dielectrics [Tho08]. This effect will be further reinforced as dielectric constants are lowered, accompanied by reductions in rigidity, in future technology nodes.

Since there is a greater risk of extrusions with low-*k* dielectrics, the void-growth-saturation effect will play an ever-declining role in future technologies.

4.6   Multiple Vias

4.6.1   Fundamentals

Vias are the elements of the interconnect that are most susceptible to manufacturing faults. Double vias are often introduced in the layout after successful routing to improve reliability (Fig. 4.20, middle).

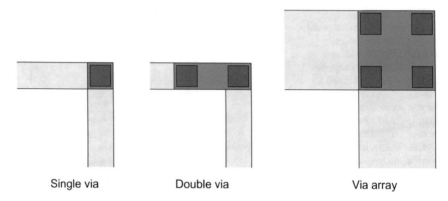

Single via   Double via   Via array

Fig. 4.20 Different types of via interconnects between metal 1 (horizontal) and metal 2 (vertical) in plan view

The reliability is increased in this way because, among other things, redundant vias are available that could independently fail, as a result of the manufacturing process, without the entire circuit failing. Redundancy allows the electrical interconnection to remain intact despite specific tolerances in the manufacturing process being exceeded. Hence, the probability of a total failure of an interconnect is reduced through the use of redundant vias.

Redundant vias and larger via arrays (Fig. 4.20, right), such as those found in power supply nets at high currents for linking wide interconnects, should be differentiated. The primary aim of the latter is to reduce the current density by increasing the cross-section, without larger via cross-sections being available in the respective layers. The impact on the current density and thus on the electromigration characteristics, as discussed below, applies to both redundant vias and via arrays.

The trade-off between multiple vias and a single via should always be considered. A balance needs to be struck between the benefits, such as a higher redundancy level and a larger interconnect cross-section, and the disadvantages due to a bigger footprint and possible problems due to reservoirs between the vias (Sect. 4.6.3).

The redundancy introduced by multi-vias renders interconnects more robust against faults caused by manufacturing tolerances. The shift between masks for neighboring layers is an example of this type of fault. The result is a mismatch between the metal structures for vias and interconnects (Fig. 4.21). In addition, mask faults can occur as well, and missing polygons or superfluous objects (particles) on the mask can result in incorrect figures in the photoresist, which, in turn, lead to missing or surplus metal. As a result, some vias may not be contacted.

Furthermore, some vias may provide no contact due to incorrect trench etching or non-uniform metal deposition. Among the possible faults are vias that are not fully opened, or vias or interconnects insufficiently filled with metal. Figure 4.22 illustrates different faults that might occur. The impact of these faults can be

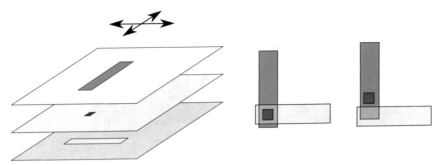

Fig. 4.21 Mask shift can cause unconnected vias; left: individual mask layers; middle: no shift; on the right: too great a shift

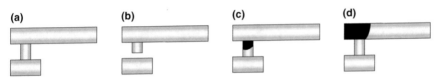

Fig. 4.22 Different faults arising from technological problems during via manufacture (side view): **a** via interconnect as designed, **b** no interconnect due to too shallow etching depth for the via, **c** insufficient metal in the via, and **d** incomplete metal deposition in the upper interconnect (black: void)

prevented by the use of redundant vias, because additional vias increase the probability that at least one contact is available.

4.6.2 Current Distribution

As explained in the preceding section, the current density can be reduced by deploying a number of parallel vias. Ideally, the current divides itself evenly between the individual vias and thus over a larger cross-sectional area. The resulting lower current density reduces the probability of electromigration damage. Another advantage is that the current running through the interconnect can be increased, as the current-carrying capacity is higher because of the enlarged via cross-section.

As depicted in Fig. 4.23, putting in place a number of parallel vias alters the individual via's current density. There is a two- to fourfold increase in the time to failure of such an interconnect with regard to electromigration.

As we shall see, these benefits can only be maximized if the geometrical configuration allows for an even current distribution (Sect. 4.6.4).

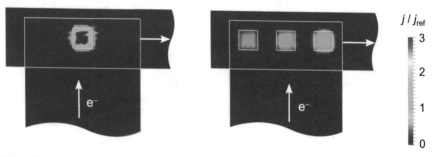

Fig. 4.23 A comparison of current densities for a single via and redundant vias

Fig. 4.24 Different types of reservoirs (highlighted in red) in an interconnect with a number of vias (gray)

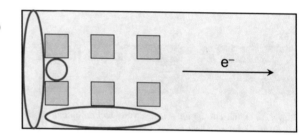

4.6.3   Vias With Reservoirs

As already noted, reservoirs can positively impact the time to failure due to electromigration and are an important consideration in the use of redundant vias [MIM+07]. When using multiple vias, reservoirs are located in the metal layers between the vias and in the overlapping sections of the connected wires (Fig. 4.24).

As described in Sect. 4.5, reservoirs change the accumulated mechanical stress produced by the material transport. If a reservoir acts as a sink for the material transport, the mechanical stresses needed for the backward transportation are reduced. Hence, the effect of reservoirs on the time to failure of the related via(s) is either positive or negative, depending on the direction of the current.

In contrast, the reservoir effect is primarily responsible for improving the time to failure with redundant vias, according to [MIM+07]. The geometrical configuration is important for assessing the effectiveness of redundant vias, as outlined next.

4.6.4   Geometrical Configuration

The geometrical configuration of redundant vias and via arrays with regard to the wires connected by them is the key property for evaluating their effect on electromigration. If the current is unevenly distributed across the redundant vias, large

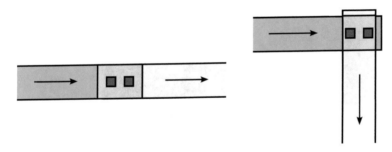

Fig. 4.25 Comparison between interconnects with and without a change in direction. The respective lengths of the current paths through the two vias differ with a change in direction; hence, more current flows through the left-hand via in the figure on the right

deviations from the mean value of the current densities in the vias occur. In extreme cases, this can overload individual vias.

In this context, the actual current density depends on the geometrical configuration of the vias in relation to the connected wires. If, for example, there is a change in direction between the interconnect layers, the vias at the "inside curve" are stressed to a greater degree than those at the outside of the curve.

This effect occurs, for example, with two parallel vias and a change in direction of the current flow (Fig. 4.25). Hence, care should be taken that the vias are so aligned that they are in the same position and orientation with respect to the angulation. The current will then be distributed evenly between the individual vias, and the current density will be reduced in an optimum manner.

How the current flows in an interconnect depends on the via configurations based on the number of redundant vias and the directions of the linked interconnects. Take, for example, a configuration of a number of vias in series along the axis of an interconnect connected orthogonally to an interconnect in a different layer. The via with the shortest current path carries the most current in this case (Fig. 4.26). The same applies to two-dimensional via arrays (Fig. 4.27).

In addition to the current-density distribution, the geometrical configuration affects the size of the reservoirs in the vicinity of the vias (Sect. 4.6.3). This reservoir size depends to a large extent on the spacing between the vias and their size. The spacing is essentially specified by the technology, in the form of minimum spacings or a fixed pitch. The maximum spacing is bounded by the permissible footprint for the via. A relationship between the via configuration and the local routing density is thus established, as redundant vias may be added only if the routing capacity is not fully exploited.

There are a number of trade-offs to be considered when selecting the via spacing (via pitch) that leads to the best utilization of reservoirs. While the size of the reservoirs grows with increased spacing between the vias, the effective rigidity of the dielectric between metal layers is reduced. Rigidity is nonetheless required to produce mechanical stresses for back diffusion (Sect. 4.3). Current distribution is also typically degraded if the via pitch it too large.

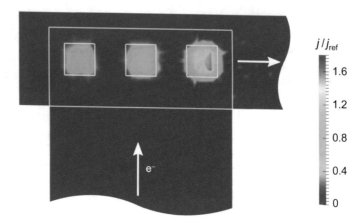

Fig. 4.26 Current densities in vias in an interconnect with many redundant vias with a change in direction between the layers; the scaling of the current density has been changed w.r.t. Fig. 4.23 for reasons of clarity

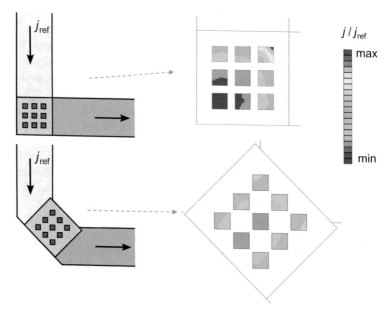

Fig. 4.27 Optimization of current-density distribution in a via array [Lie05, Lie06]

4.6.5 Applications

The effect of redundant vias on electromigration is still an active research area. The results of studies, such as [DOC08, RT08, CKY+11], show that the use of redundant vias has a positive impact on the time to failure in the presence of

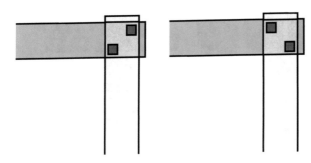

Fig. 4.28 A comparison of different via configurations for a change in direction. In the left-hand figure, the bottom-left via is stressed to a greater degree than the via above it to the right. The vias are evenly stressed in the figure on the right

electromigration stress in power supply nets especially. Two issues need to be addressed w.r.t. their use in signal nets. Do their benefits predominate when the formation of reservoirs is taken into consideration? And is there enough space available?

Double (redundant) vias are utilized today in IC design as standard in "post-layout techniques". The main driver behind their use is the increased yield during manufacturing. Vias are added in as many locations as possible in the layout as long as the layout is not enlarged and not altered topologically. Other boundary effects, such as the current-density distribution and reservoirs, are mostly ignored at the present time.

Via arrays are introduced in power supply nets, especially in areas where high currents are expected. The current density and the static voltage drop are crucial in this regard. Initial attempts are being made to optimize the via configuration for the best possible current distribution [Lie06, BL16].

The change in electromigration characteristics due to current distribution and the occurrence of reservoirs when redundant vias are introduced must be considered. Both of these side effects depend on the geometrical configuration of the vias and the connected wires (Fig. 4.28). Simulations, such as depicted in Fig. 4.26, show that the vias should be placed with minimum spacing on the perpendicular to the bisecting line between the connected wires.

Ideally, the two connected wires run in the same direction (and thus have the same orientation), so that all possible current paths have the same length. This can be achieved by routing one of the connected wires "around the corner" (Fig. 4.29). This can only be done, however, if the routing layout is not forced to follow preferred directions.

Special attention should be given to the (sometimes unintended) formation of reservoirs with EM-compliant redundant vias, as they may degrade the expected gain on the time to failure [MIM+07].

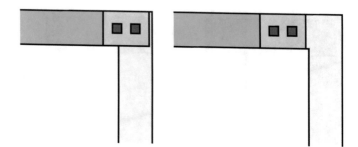

Fig. 4.29 Two solutions for the problem of non-uniform via stressing with a change in direction of the wiring. By placing the via array "in line" with the current direction, all possible current paths have the same length, thereby ensuring a uniform current distribution between the two vias

4.7   Frequency-Dependent Effects

4.7.1   Self-Healing and Rising Frequencies

As explained in Sect. 2.4.3 (Chap. 2), the vast majority of nets in digital circuits carry currents for which self-healing plays an important role.

The self-healing effect can be included in models for calculating an equivalent current density. A principle that enables this is represented by the forward and backward diffusion flows J in Eq. (4.10) as follows:

$$J_{net} = J_{forward} - J_{backward} = J_{forward} \cdot (1 - \gamma), \qquad (4.10)$$

where γ is the self-healing coefficient [TCH93, TCC96].

As outlined in Sect. 3.3.1 (Chap. 3), using the value of the average current (and not the RMS current value) can also account for the consideration of the self-healing effect.

At present, there is a specific range for clock and signal frequencies in integrated circuits within which self-healing in signal lines is effective (Fig. 4.30). This range, however, does not include the skin effect (yet). Thus, the time to failure in current frequency ranges is much greater than for DC currents. However, as is evident from Fig. 4.30, the reliability of the interconnects cannot be increased by scaling the frequency any further. Self-healing is negligible in power supply nets, and in signal lines with low data rates, as illustrated in the left-most part of Fig. 4.30.

With the skin effect, the current is "forced" into the exterior of the interconnect. This impacts the current density and the time to failure at very high frequencies (Sect. 2.4.3, Chap. 2). The frequencies in question are in the terahertz range for presently used interconnect dimensions. Clock frequencies at present reach a maximum of 6 GHz, which is significantly lower than the frequency that is critical for the skin effect (see Fig. 4.30). The skin effect is active only with the high-frequency components that are driven with very steep signal edges.

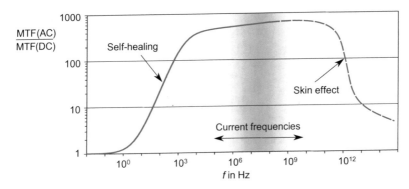

Fig. 4.30 Frequency dependency of the median time to failure, according to [TCH93, WY02]. The time to failure rises in the kilohertz range due to self-healing and falls in the terahertz range due to the skin effect (Sect. 2.4.3, Chap. 2)

The time to failure of interconnects is practically unaffected by the frequency f in presently used switching-frequency ranges (plateau range in Fig. 4.30). At the same time, thermal migration has a significant influence on the time to failure in the higher frequency ranges. The EM-driven mass transport, which is dependent on the direction of current flow, has less of an influence here, while heating remains constant for a constant current RMS value. This means that the thermal-migration effect rises at high frequencies (Sect. 2.5, Chap. 2).

4.7.2 Applications

Nets in digital electronic circuits should be categorized according to the dependence of their time to failure on the frequency.

Power supply nets are classed as the most sensitive to EM, as they have the lowest (or no) frequencies coupled with high current values. Clock and signal nets are put in another net class as they have high frequencies and thus self-healing attributes. Within this net class, clock and signal nets differ in terms of their current values and the symmetry of their current characteristics.

In an attempt to improve the time to failure, especially in the case of power supply nets, direct current (DC) nets could be replaced with alternating current (AC) nets, to take advantage of the self-healing effect. However, only global power supply nets across the entire IC can be replaced in this way, as DC nets are still required locally. This solution is practicable only for special applications with very high currents due to the considerable amount of auxiliary circuitry required and additional electromagnetic-coupling problems.

The main conclusion to be drawn from the frequency analysis is that the use of a global current-density boundary for all nets is not beneficial. We therefore recommend different net classes, each with respectively different boundaries,

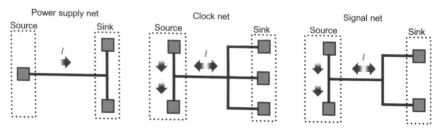

Fig. 4.31 Three net classes with their major current and topological characteristics

which can be derived from an analysis of the expected currents and frequencies. Not taking bus systems into account, there are three main net classes: power supply nets, clock nets, and signal nets (Fig. 4.31).

Power supply nets are the most sensitive to EM, as they typically carry DC currents. AC currents predominate in clock and signal nets, except for small areas within the signal source. The main difference between these two net classes (clock and signal) is the number of sinks per net. This number is usually higher in clock nets, where it also leads to higher currents.

Figure 4.30 shows that it is beneficial to introduce at least two different boundary values for the current density and hence, the EM stress: one for DC currents or frequencies below 10 kHz and one for AC currents with higher frequencies. In the case of DC currents or frequencies below 10 kHz, further boundary values can be defined for the transition between 10 Hz and 10 kHz.

4.8 Materials for Classical Metal Routing

All materials in the routing layers of an IC are critical for electromigration. These include the interconnect materials in the metallic wires and metallic via fillings (vias), dielectrics in both types of layers, and barriers and adhesive layers between metal and metal, and between metal and dielectric. The different domains (metal, dielectric, and barriers) are shown schematically in Fig. 4.32.

Fig. 4.32 Sectional view of a copper interconnect with the necessary barrier layers (dielectric cap and metal liner) and its surounding dielectric (not to scale)

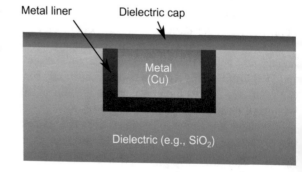

The materials in the interconnect, the dielectric, and materials in the barriers can be modified to improve the electromigration characteristics. The available options for all three domains are described in the following Sects. 4.8.1–4.8.3.

The barriers are especially crucial for electromigration (see also Chap. 2), as they form the boundary layer—a particularly EM-sensitive region with copper metallization. The purpose of barriers in terms of electromigration is to seal open surfaces and to increase the activation energy for the diffusion at the boundary layers between copper and dielectric.

One must take into account that if a particular diffusion mechanism is inhibited, another mechanism will always become dominant, leading to large variations in damage. For example, although the switch from aluminum to copper blocked grain-boundary diffusion, it promoted surface diffusion. Furthermore, if surface diffusion is suppressed with appropriate dielectric and barrier layers, the grain-boundary diffusion resurfaces as the dominant type of diffusion. Bulk diffusion may come to dominate if all other mechanisms are disabled. Hence, it is not enough to examine the barrier materials here on their own.

4.8.1   Interconnect

The choice of interconnect material has the greatest impact on electromigration, as the process takes place within the interconnect material itself. The interconnect material must have the following properties:

- low specific resistance,
- good EM robustness, and
- compatibility with semiconductor processes.

In the late 1990s, the wiring material was changed from aluminum to copper due to the first property. The lower specific resistance of copper resulted in a lower time constant $\tau = RC$, which gave rise to higher signal frequencies. This, in turn, caused a lower voltage drop and lower self-heating in the interconnects. The changeover boosted overall electromigration robustness.

Metallization-technology changes were also needed to ensure compatibility with the semiconductor processes and materials. This gave rise to the *Damascene* technique (Sect. 2.4.2), as the direct lithographical structuring of copper was more complex than with aluminum.

Metals can be doped with other substances to make them more resistant to electromigration damage (this topic is dealt with in more detail in Sect. 4.8.3). This technique has proved successful with copper-doped aluminum wiring, where the diffusion paths at the grain boundaries are blocked by copper.

The properties mentioned at the outset must be kept in mind when advocating for a metal change. Table 4.2 clearly shows that copper with its low specific resistance

Table 4.2 Specific resistance and activation energy for the migration of voids in metals, resistance values from [BF83]

Metal	Specific resistance ϱ in $\mu\Omega \cdot cm$	Activation energy E_a in eV	Source
Aluminum	2.44	0.61	[EJS91]
Silver	1.47	0.66	[EJS91]
Copper	1.54	0.70	[EJS91]
Gold	2.03	0.75	[EJS91]
Tungsten	4.84	1.89	[MIH+90, EJS91]

outperforms other elemental metals, which we discuss further below. Deploying other materials would mean an increase in the resistance or a decrease in the activation energy.

Silver has the best conductivity of all known elemental metals; however, it has a low activation energy and behaves poorly as compared to copper in terms of electromigration and its effects. Gold has favorable conductivity, but unfortunately acts as a poison to semiconductors: it destroys the semiconducting properties of silicon by adding energy levels. To address this, the metal liners would have to be significantly upgraded. Despite finding use as contacts between interconnects and polysilicon and having a high EM robustness [QFX+99], tungsten has not been deployed in other applications due to its high resistance.

Switching from metals to new types of materials, such as carbon-based nano-materials, is a possible option. We will discuss these new materials in more detail in Sect. 4.9. Let us first elaborate on classical metallization in this Sect. 4.8. In addition to the aforementioned interconnect material, the interconnect environment, such as the dielectric (Sect. 4.8.2) and the barrier (Sect. 4.8.3), also affects the EM characteristics, as described below.

4.8.2 Dielectric

The dielectric has a bearing on electromigration by means of mechanical boundary conditions. As explained in Sect. 4.3, mechanical stresses cause stress migration, which can beneficially slow down or prevent void growth. The scale of the mechanical stress and thus the magnitude of the back diffusion as well depends on the rigidity of the interconnect surroundings. For a given geometrical configuration, this rigidity is determined to a large extent by the Young's modulus of the dielectric.

Silicon oxide (SiO_2), with its very good technological properties, is notable for its fairly high Young's modulus (Table 4.3) and is therefore suitable for enabling back diffusion. Hence, the Blech length is quite large due to the stress migration.

Table 4.3 Dielectric
constants ε_r (or k) and
elasticity moduli E for
important insulating materials

Material	ε_r	E in GPa
Vacuum or air	1	–
Aerogels	1.1–2.2	≈ 0.001
Polyimide (organic)	2.7	3.7
Silicon oxide (SiO$_2$)	3.9	300
Glass fiber-reinforced epoxy resin	5	20
Silicon nitride (Si$_3$N$_4$)	7.5	297
Aluminum oxide (Al$_2$O$_3$)	9.5	264
Silicon	11.7	99

There are nonetheless reasons for changing the dielectric:

- The time constant $\tau = RC$ must be reduced for smaller structures in order to increase the frequencies. A lower dielectric constant is beneficial in this respect (low k).
- The capacitance C must be increased without reducing the spacing. A higher dielectric constant is required in this scenario (high k).

The second option impacts only the gate for field-effect transistors. Specifically, high-k materials help lower the power dissipation of the IC, as a high transistor gate capacitance correlates with a low resistance in the ON state.

Low-k materials are required in the routing as an insulator between the metal layers, so-called interlayer dielectric (ILD), to reduce the capacitances between different interconnects and thus to increase the permitted signal frequencies. Two different approaches, that can also be combined, are available for this purpose:

- Parts of the dielectric are replaced with air or with a vacuum. This is achieved with a porous dielectric or an aerogel.
- Alternative materials, such as organic materials with a lower dielectric constant, are used.

The use of both variants means that the effective Young's modulus of the dielectric is lowered and that, with porous materials especially, there is a higher risk of metal extrusions penetrating the dielectric. The best results are achieved by combining the two techniques, thus obtaining a trade-off between dielectric constant and Young's modulus.

For these reasons, the specific dielectric must be optimized for its particular application. An attempt should be made to choose dielectrics with a high Young's modulus, despite a lower dielectric constant. Porous SiCOH (carbon-doped oxide dielectrics comprised of Si, C, O, and H) and different organic dielectrics are good options [ITR14, ITR16]. There are certain compensatory measures that can be taken to increase rigidity: the dielectric can be modified locally or metal structures can be supplemented [Tho08].

In other cases, the boundary values for current densities (current-density limits) should be modified such that the lower back diffusion is respected.

4.8.3 Barrier

As mentioned before, a barrier is required to prevent diffusion between conductor and dielectric. Without a barrier, the metal and dielectric properties would be altered too much by the high temperatures encountered during manufacture [UON+96].

Key barrier attributes are:

- good adhesion to copper and the dielectric, and
- high thermal and mechanical stability for a low coating thickness (few nanometers).

The most popular metallic liners are made of tantalum (Ta) [MGL+11, OLM+01, HOG+12], titanium (Ti) [DST95, LTC98, OLM+01], and compounds of these metals, like tantalum nitride (TaN) or titanium nitride (TiN). The metal liner can easily be deposited in thin layers in trenches etched for the interconnects. Due to the metallic properties, adhesion to the interconnect material is good, and some conductance capability remains if the interconnect is broken. This type of barrier layer is in place as well at the interface between the via and underlying metal layer; the barrier layer must therefore conduct the current.

The metal liner ensures quite high activation energies at the copper boundary with regard to electromigration (Table 4.4).

A barrier (in the form of a dielectric cap) to the superior dielectric must be put in place after the material abrasion in the CMP step. The dielectric cap is typically composed of a dielectric material. Hence, the cap does not require any further structural changes; it only needs to be opened later at the vias. This operation is carried out when the trenches are etched. Silicon and nitrogenous compounds are the materials of choice here, with compounds silicon nitride (SiN) [OLM+01, MGL+11], silicon carbon nitride (SiCN) [MGL+11], and hydrogen silicon carbon nitride (SiCNH) being good examples of dielectrical caps. The first of these compounds, silicon nitride, is known for its compatibility with semiconductor processes, as it was deployed in semiconductor manufacture as a hard mask, etch-stop, or dielectric prior to its use as a dielectric cap.

Table 4.4 Activation energies for the boundary layer between copper and different barrier materials

Barrier	Activation energy E_a in eV	Sources
Ta	2.1	[Gla05]
Ta/TaN	1.4	[HGR06]
SiN or $SiC_xN_yH_z$	0.7–1.1	[Gla05, HGR06]
SiN on Cu(Ti)	1.3	[HGR06]
CoWP	1.9–2.4	[HGR06]
SiC_xH_y	0.9	[HGR+03]

The dielectric cap requires different properties than the metal liner. It must adhere well to the metal and the dielectric, in order to support other routing layers on top of the dielectric cap.

Dielectric materials, however, adhere worse than metals to copper. In addition, the top surface of the interconnects is quite flawed due to CMP. This is why the activation energy at the boundary layer between the upper copper surface and a dielectric cap is low (see third row in Table 4.4.).

This problem can be avoided by using metallic barriers, for example, tungsten (W), tungsten nitride (WN, W_2N) [UON+96], or cobalt tungsten phosphide (CoWP; see Table 4.4) [HGR+03] on the top surface of the interconnect. However, if metallic barriers are used, the barrier structure must be adapted to fit the interconnect layout to avoid short circuits.

This structuring can be performed in (i) an additional lithographical step in a similar manner as with the interconnect layer underneath or (ii) a self-aligning barrier [CLJ08, LG09, VGH+12]. A self-aligning barrier is preferable to an extra lithographical step, as the latter is more expensive and may produce more defects. The barrier must accumulate exclusively on the copper surface and recess the exposed dielectric.

The easiest way to produce this type of barrier is to dope the interconnect material with the appropriate barrier material. The material is then annealed, whereby the barrier material diffuses to the boundary layer of the interconnect material to produce the barrier [LG09, VGH+12]. Only some barrier materials, like manganese (Mn) [YML+11, HOG+12] or ruthenium (Ru) [YML+11], are suitable for this treatment.

Barriers pose additional technological challenges, especially with further reductions in structure sizes. Lithographically manufactured structures have now reached sizes in the lower nanometer range, which means that barriers must have thicknesses in the sub-nm range. This causes problems, as there are only a few atomic layers available for the barrier function. For example, the barrier must be stable over the longer term to function properly. In addition, there should be no diffusion between barrier and copper or barrier and dielectric, which would inadmissibly modify the barrier properties or the environment.

The effort required to introduce the above chemical elements varies greatly. Depending on how the materials must be treated, they can either be quite easily integrated in the manufacturing process or may require modification of the entire process. Similarly, increased thermal stresses caused by depositing certain barrier materials must be compensated with appropriate measures. Hence, for example, the thermal stress in other manufacturing steps must be restricted, as there is only a limited thermal budget available for semiconductor structures during manufacture.

Many materials serve as barriers in practice. The high degree of dependence between technology parameters and structural parameters on the deployed material combination should always be accounted for when considering electromigration effects.

4.9   New Materials and Technologies

The use of different interconnect materials is one way to suppress EM in the presence of ever-decreasing semiconductor scale and increasing current densities. As discussed in Sect. 4.8, the prospect appears remote for finding in the near future a suitable metal with higher EM robustness than the currently employed metal, copper. In the following sections, we will describe new solutions that go beyond classical metal routing.

4.9.1   Carbon-Based Solutions

Due to their outstanding physical properties, carbon-based materials are proposed for future interconnect technology. Three possible approaches are being researched at this time, involving different carbon allotropes: the interconnect may be realized by an array of *graphene layers* (graphene nanoribbons, GNRs), by *carbon nano-fibers* (CNFs), or by a bundle of *carbon nanotubes* (CNTs).

Graphene is a sheet of carbon atoms arranged in a 2D honeycomb lattice (Fig. 4.33), which may be further stacked or arranged. For example, carbon nanofibers are cylindric nanostructures with graphene layers arranged as stacked cones, cups, or plates. In one method, carbon nanotubes are obtained by rolling up a graphene sheet; carbon nanofibers with graphene layers wrapped into perfect cylinders are also called carbon nanotubes.

Carbon nanotubes (CNTs) are one of the most promising solutions for electrical interconnects in nanoscale structures. Many publications describe CNT as the interconnect material of the future [BS06, SB04]. Reliability in highly integrated electronic circuits is cited as being a key consideration [CSX+11]. CNT compares very well with copper w.r.t. reliability [AKH+09].

Carbon nanotubes are composed entirely of sp^2 bonds,[1] similar to graphite, which is the most stable form of carbon under standard conditions. Note that sp^2 bonds are stronger than the sp^3 bonds found in diamond. This bonding structure provides carbon nanotubes with their unique strength. Moreover, they align themselves into ropes held together by the van der Waals force and can merge together under high pressure. When merging, they trade some sp^2 bonds for sp^3 bonds and produce very strong wires of a nano- to micrometer lateral dimension.

---

[1]When carbon bonds, the shape of the "p" orbitals will change to a different shape to allow for less repulsion between electrons. For a carbon with one double bond and two single bonds, the orbitals will become 33% "s" and 67% "p", making it "sp^2". Hence, the term "sp^2" indicates that one S shell mixes with two P shells.

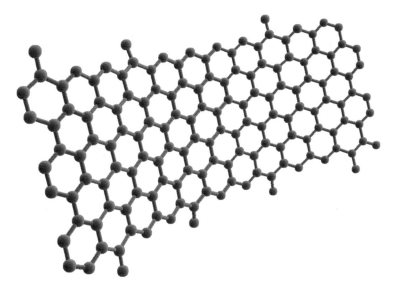

Fig. 4.33 Schematic view of a graphene structure

CNTs consist of one or more graphene layers that one can imagine being rolled up to form a tube. Thus, the following types of CNT are to be found:

- single-wall CNT (SWCNT), and
- multi-wall CNT (MWCNT).

The form of nanotubes is identified by a sequence of two numbers. The first one represents the number of carbon atoms around the tube, while the second number identifies an offset of where the nanotube wraps around to. The electrical properties depend on the orientation of the carbon layers in the tubes.

Single-wall CNTs can be divided into three groups, according to their chiral vector or chirality, which is visible in the trajectory of the atomic connections along tube circumferences (Fig. 4.34):

- armchair,
- zigzag, and
- chiral.

Depending on the orientation, there are semiconducting and metallically conducting CNTs, whereby one-third of the possible orientations is metallically conducting. Multi-wall CNTs are almost always metallically conductive, because it is highly likely that at least one tube wall is conductive.

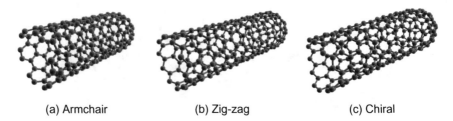

(a) Armchair   (b) Zig-zag   (c) Chiral

Fig. 4.34 Schematic views of various single-wall CNTs, each with different atomic-connection patterns

Three-dimensional structures (crystal lattice) and one-dimensional structures, such as nanotubes, have been well known for some time, but it is only recently that inherently two-dimensional structures have emerged with the discovery of graphene.

Graphene consists of the individual atomic layers of graphite (see Fig. 4.33) in which the carbon atoms form a planar hexagonal lattice. This structure has extraordinary mechanical and optical properties. The electrical resistance of graphene in the lateral direction is very low [ST10]. IC routing connections can benefit from this property: depending on the structure sizes, interconnects are either composed of individual graphene platelets or structured as graphene layers.

Carbon nanofibers (CNFs) are elongated objects similar to CNTs. They consist of stacked graphene layers arranged as plates or cones. CNFs, hence, are also prospective candidates for interconnects. CNTs can thus be treated as a special case of CNFs, where graphene layers run parallel to the longitudinal axis of the fibers and the conical shape becomes cylindrical.

As CNTs have very good conductivity and are readily available, we will describe them in more detail below as an example of the different carbon modifications. They are available in many variations, and practice-oriented research results are on file, too.

4.9.2   CNT Properties

Metallic CNTs have both high thermal and electrical conductivity, while also providing thermal stability, which is due to the strong sp^2 bonding between carbon atoms. CNTs are much less susceptible to EM than copper interconnects and can carry high current densities. Values of up to 10^{10} A/cm^2 for single CNTs have been measured thus far [BS06]. Ballistic electronic transport can go up to 1 μm nanotube lengths, enabling CNTs to carry very high currents with virtually no heating; this is due to the nearly 1D electronic structure [TRO17].

A copper interconnect with a 100×50 nm cross-section can transfer currents up to 50 µA, whereas a CNT with a diameter of 1 nm can carry currents up to 20–25 µA [PRY04]. Hence, a few CNTs can (theoretically) match the current-carrying capacity of a much larger copper interconnect.

In addition, the thermal conductivity of CNTs ranges from 3000 to 10,000 W/(m·K) at room temperature [MPG13], thus, exceeding that of copper [385 W/(m·K)] by approximately a factor of 10. Hence, heat diffuses much more efficiently through CNTs. This supports their use not only in thermal vias but also in through-silicon vias (TSVs) for 3D integration, where (vertical) thermal conduction is urgently required.

The properties of single-wall (SWCNT), multi-wall (MWCNT) CNTs and Cu-CNT composites, where a mixture of CNTs and copper is used, are summarized in Table 4.5.

Obviously, CNTs are in theory superior to copper for all relevant interconnects properties. However, challenges arise with their use, which will require different technological processes in some areas. In addition, many of the favorable properties of real-world CNT structures are significantly less than the theoretical limits noted earlier. For example, the theoretical electrical resistance of individual CNTs increases greatly in CNT arrays due to the contact resistance [LYC+03, CSX+11].

Figure 4.35 shows the theoretical resistivity comparison for copper wires and single- and multi-wall CNTs for different lengths and diameters up to 80 nm.

Table 4.5 Comparison of properties of Cu and CNTs, indicating the enormous potential of CNTs to serve as future, EM-robust interconnects

	Cu	Single-wall CNTs	Multi-wall CNTs	Cu-CNT composites
Maximum current density (A/cm^2)	$<1 \times 10^7$ [TRO17]	$>1 \times 10^9$ [YKD00]	$>1 \times 10^9$ [WVA01]	$>6 \times 10^8$ [SSY16]
Thermal conductivity @300 K [W/(m·K)]	385 [TRO17]	3000–10,000 [MPG13]	3000 [MPG13]	~ 800 [SYK13]
Electrical conductivity (S/cm)	5.8×10^5 [TRO17]	7×10^5 [LYY04]	2.7×10^5 [KKR12]	$(2.3–4.7) \times 10^5$ [SSY16]
Electron mean free path @300 K (nm)	40 [TRO17]	>1000 [PHR07]	$>25,000$ [BYW02]	
Coefficient of thermal expansion (1/K)	17×10^{-6} [BS06]	$(-0.3–+0.4) \times 10^{-6}$ [JLH04][a]		$(4–5) \times 10^{-6}$ [ARW11]
Young's modulus (GPa)	129 [Bei03]	~ 1000 [Bel05, MR06]	~ 900 [MR06]	
Tensile strength (GPa)	0.2 [BS06]	~ 100 [MR06]		

[a]@400 K, axial direction; the specific value depends on temperature, diameter, CNT structure, and direction

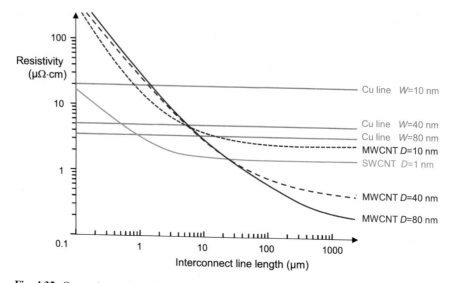

Fig. 4.35 Comparison of electrical resistivity between copper lines (extrapolated from ITRS [ITR14]) and single- and multi-wall CNTs (based on analytical models [CGJ14]) for different interconnect lengths (x-axis), Cu widths (*W*), and CNT diameters (*D*), according to [TRO17]

Clearly, the theoretical resistivity of multi-wall CNTs for larger lengths (>10 μm) is not just significantly smaller than that of copper but outperforms single-wall CNTs, as well. Please note, however, that measured resistivity values for MWCNTs are still considerably higher than those predicted in these analytical models [CGJ14].

CNTs have a low thermal expansion coefficient combined with a high mechanical tensile stressability [MR06]. This attribute can be leveraged to compensate for differences in thermal expansion. For example, by manufacturing composites of copper and CNT [ARW11], their thermal coefficient of expansion can be adjusted to that of silicon or silicon oxide. The addition of CNT is being investigated for this reason in other fields as well, such as filler metals for contacting integrated electronic circuits to substrates [XZK+12].

4.9.3 Applications

Carbon nanotubes can be deployed in layout design in the form of single CNTs, Cu-CNT composites [CC08, Cha09], and CNT arrays. The latter are vertically aligned carbon nanotube arrays consisting of carbon nanotubes oriented along their longitudinal axes normal to a substrate surface. New routing techniques are, to some degree, needed for these arrays.

While standard technologies, such as metal deposition and CMP, are typically employed for Cu-CNT composites, single CNTs and CNT arrays require new technological processes. CNT arrays are especially suited for vertical interconnects (vias). Here, a number of process flows have been developed to selectively grow densely packed bundles of CNTs on the metallic bottom electrode and to contact the other end of the CNTs with the top metal. Chemical vapor deposition (CVD) is used for this purpose; CVD is generally performed at very high temperatures. Processes have not yet been developed to create CNTs at low temperatures. The interconnects will continue to be made of metal or a composite material, while the application of CNT arrays is pursued.

Connecting CNTs to their surroundings and to other structures, as well as between one another, is challenging. There is a risk of high contact or transient resistance with metallic connections, which would neutralize the benefits of low line resistance and high current-carrying capacity that are being sought. Doped CNTs and different functionalizations and liners are under research at this time to address these issues [CHT+12, SBK05]. Another problem arises due to the fact that the current should be injected ideally alongside the axis of the CNT to create a so-called end-bonded contact to the CNTs [TRO17]. This, however, requires perfect control over the quality of the interface between open-ended CNTs and metal.

In addition to the contact issues, the electrical performances of CNTs integrated as interconnects are still lower than those predicted in theoretical models, and even lower than that of copper. This can be explained by the insufficient packing density of CNTs in interconnects and the quality of the CNTs produced by the low-temperature CVD process [TRO17].

Due to these technological challenges, interconnects made of CNTs are unlikely to replace all copper interconnects at once. Rather, hybrid structures, consisting of vertical parallel CNTs for vias combined with interconnects made of composite materials, are a possible solution.

Interconnects are difficult to produce with aligned CNTs at this time. However, unordered CNTs embedded in metal allow for a compensation of the thermal expansion. Yet high conductivity and robustness against electromigration can only be achieved with aligned CNTs. Hence, the focus is now on so-called *aligned CNT-Cu composites*, where the CNT alignment in the composite is controlled by current flow.

Theoretically, there can be no electromigration within CNTs, as the activation energy or the bonding energy of at least 3.6 eV is very high, as compared to the activation energy of copper of approximately 1 eV. However, this useful characteristic comes with a severe drawback: every void in the CNT greatly impedes the ballistic current flow; this leads to an extreme rise in self-heating. Thus, the CNT is immediately destroyed if the maximum permissible current (density) is exceeded.

We have seen that carbon-based solutions hold great promise, with highly favorable electrical, thermal, and physical properties. It is the challenge of the next generation of researchers and engineers to make such solutions viable in practice.

4.10   Summary

In this chapter, we examined different methodologies for mitigating electromigration in integrated electronic circuits. The use of the bamboo effect, length and reservoir effects, via configurations, self-healing, and the use of appropriate materials for conductors, dielectrics, and barriers are some of the issues we have examined.

The bamboo effect can greatly prolong the time to failure of an interconnect; it does, however, depend on numerous constraints. Before the effect can work in copper interconnects, the surface diffusion must be disabled, which requires specific materials. The focus in this regard is on very stable barrier layers that raise the activation energies to a high level. The interconnect resistance is increased as well in massively miniaturized technologies by means of wire slotting due to scattering effects. The benefits of the bamboo structure are thus largely eliminated by the negative thermal effects of self-heating.

The critical length effects in combination with via effects (via-below and via-above configurations) and reservoir effects will soon become beneficial. Reliability can be greatly improved by controlling the wiring by using short segments where possible, and considering different boundary values for the product of current density and length. Reservoirs should be deployed in appropriate applications, as well.

Special care should be taken when arranging redundant vias or via arrays to avoid generating new EM flash points in the layout. Via arrays are required in the very EM-sensitive power-supply nets. Hence, they are critical for the digital design flow. Specifically, you should optimize the via configurations also with regard to reservoirs between the vias to maximize the time to failure for small footprints. Reservoirs are particularly beneficial in power-supply nets due to the constant direction of the current flow.

Frequency effects, that is, self-healing and skin effects, serve only to classify nets and allocate current-density boundary values. This is critical for the routing step and also for verification, as different current-density boundary values should be assigned for each net class. Valuable routing resources would be squandered by oversizing if a global boundary value was applied to the entire layout. The skin effect will not play a significant role in digital circuits now and in the near future, because frequencies will only increase very slowly compared to the scaling of the structural miniaturization. In future, on the other hand, the skin effect will affect reliability in analog high-frequency integrated circuits for frequencies of 45 GHz or higher [YZZ+11].

In this chapter, we also cast an eye beyond current methodologies and into the future. The main focus of present research is on the search for a new interconnect material, possibly CNTs. The current state of research shows that CNTs are potential candidates as a via and wire material. Further practice-oriented research is required, however, to integrate them in the manufacturing process for integrated circuits.

References

[AKH+09]   N. Alam, A.K. Kureshi, M. Hasan, et al., Analysis of carbon nanotube interconnects and their comparison with Cu interconnects, in *IMPACT '09* (2009), pp. 124–127. https://doi.org/10.1109/mspct.2009.5164190

[Aro03]   N.D. Arora, Modeling and characterization of copper interconnects for SoC Design, in *International Conference on Simulation of Semiconductor Processes and Devices (SISPAD)* (2003), pp. 1–6. https://doi.org/10.1109/sispad.2003.1233622

[ARW11]   L. Aryasomayajula, R. Rieske, K.-J. Wolter, Application of copper-carbon nanotubes composite in packaging interconnects, in *34th International Spring Seminar on Electronics Technology (ISSE)* (2011), pp. 531–536. https://doi.org/10.1109/isse.2011.6053943

[Bei03]   P. Beiss, Non-ferrous materials, in *Powder Metallurgy Data*, ed. by P. Beiss, R. Ruthardt, H. Warlimont. Landolt-Börnstein—Group VIII Advanced Materials and Technologies, vol. 2A1 (Springer, 2003), pp. 460–470. https://doi.org/10.1007/10689123_23

[Bel05]   S. Bellucci, Carbon nanotubes: physics and applications. Phys. Status Solidi (c), **2**(1), 34–47 (2005). https://doi.org/10.1002/pssc.200460105

[BF83]   J. Bass, K.H. Fischer, Metals: electronic transport phenomena, in *Landolt-Börnstein—Numerical Data and Functional Relationships in Science and Technology*, ed. by J.L. Olsen, K.-H. Hellwege. Group III Condensed Matter, vol. 15A (Springer, 1983). https://doi.org/10.1007/b29240

[BL16]   S. Bigalke, J. Lienig, Load-aware redundant via insertion for electromigration avoidance, in *Proceedings of the International Symposium on Physical Design (ISPD 2016)*, pp. 99–106. https://doi.org/10.1145/2872334.2872355

[Ble76]   I.A. Blech, Electromigration in thin aluminum films on titanium nitride. J. Appl. Phys. **47**(4), 1203–1208 (1976). https://doi.org/10.1063/1.322842

[BS06]   K. Banerjee, N. Srivastava, Are carbon nanotubes the future of VLSI intercon- nections? in *Proceedings of the Design Automation Conference (DAC)* (2006), pp. 809–814. https://doi.org/10.1145/1146909.1147116

[BYW02]   C. Berger, Y. Yi, Z.L. Wang, et al., Multiwalled carbon nanotubes are ballistic conductors at room temperature. Appl. Phys. A **74**, 363 (2002). https://doi.org/10.1007/s003390201279

[CC08]   Y. Chai, P.C.H. Chan, High electromigration-resistant copper/carbon nanotube composite for interconnect application, in *2008 IEEE International Electron Devices Meeting (IEDM)* (2008), pp. 1–4. https://doi.org/10.1109/iedm.2008.4796764

[Cha09]   Y. Chai, *Fabrication and Characterization of Carbon Nanotubes for Interconnect Applications*. Ph.D. thesis, Hong Kong University of Science and Technology, Hong Kong (2009)

[CHT+12]   Y. Chai, A. Hazeghi, K. Takei, et al., Low-resistance electrical contact to carbon nanotubes with graphitic interfacial layer. IEEE Trans. Electron Devices **59**(1), 12–19 (2012). https://doi.org/10.1109/TED.2011.2170216

[CKY+11]   T.-F. Chang, T.-C. Kan, S.-H. Yang, et al., Enhanced redundant via insertion with multi-via mechanisms, in *IEEE Computer Society Annual Symposium on VLSI (ISVLSI)* (2011), pp. 218–223. https://doi.org/10.1109/isvlsi.2011.50

[CLJ08]   H. Chang, Y.-C. Lu, S.-M. Jang, Self-aligned dielectric cap. U.S. Patent App. 11/747,105 (2008)

[CS11]   H. Ceric, S. Selberherr, Electromigration in submicron interconnect features of integrated circuits. Mater. Sci. Eng.: R: Rep. **71**(5–6), 53–86 (2011). https://doi.org/10.1016/j.mser.2010.09.001

[CSX+11]   Y. Chai, M. Sun, Z. Xiao, et al., Pursuit of future interconnect technology with aligned carbon nanotube arrays. IEEE Nanotechnol. Mag. **5**(1), 22–26 (2011). https://doi.org/10.1109/MNANO.2010.939831

[CCT+06]   C.W. Chang, Z.-S. Choi, C.V. Thompson, et al., Electromigration resistance in a short three-contact interconnect tree. J. Appl. Phys. **99**(9), 094505 (2006). https:// doi.org/10.1063/1.2196114

[CGJ14]   J.S. Clarke, C. George, C. Jezewski, et al., Process technology scaling in an increasingly interconnect dominated world, in *Symposium on VLSI Technology Digest of Technical Papers* (2014), pp. 142–143. https://doi.org/10.1109/VLSIT. 2014.6894407

[DDM96]   J.A. Davis, V.K. De, J.D. Meindl, A priori wiring estimations and optimal multilevel wiring networks for portable ULSI systems, in *Proceedings of the Electronic Components and Technology Conference* (1996), pp. 1002–1008. https://doi.org/10.1109/ectc.1996.550516

[DOC08]   R.L. de Orio, H. Ceric, S. Carniello, et al., Analysis of electromigration in redundant vias, in *International Conference on Simulation of Semiconductor Processes and Devices (SISPAD)* (2008), pp. 237–240. https://doi.org/10.1109/ sispad.2008.4648281

[DST95]   I. De Munari, A. Scorzoni, F. Tamarri, et al., Activation energy in the early stage of electromigration in Al–1% Si/TiN/Ti bamboo lines. Semicond. Sci. Technol. **10**, 255–259 (1995). https://doi.org/10.1088/0268-1242/10/3/004

[EJS91]   P. Ehrhart, P. Jung, H. Schultz, et al., Atomic defects in metals, in *Landolt-Börnstein, Numerical Data and Functional Relationships in Science and Technology*, ed. by H. Ullmaier. Group III: Crystal and Solid State Physics, vol. 25 (Springer, 1991). https://doi.org/10.1007/b37800

[Gla05]   A. von Glasow, *Zuverlässigkeitsaspekte von Kupfermetallisierungen in Integrierten Schaltungen*. Ph.D. thesis, TU Munich (2005)

[Gup09]   T. Gupta, *Copper Interconnect Technology* (Springer, 2009). https://doi.org/10. 1007/978-1-4419-0076-0

[HC08]   M.S. Hefeida, M.H. Chowdhury, Interconnect wire length estimation for stochastic wiring distributions, in *International Conference on Microelectronics* (2008), pp. 369–372. https://doi.org/10.1109/icm.2008.5393767

[HGR+03]   C.-K. Hu, L. Gignac, R. Rosenberg, et al., Reduced Cu interface diffusion by CoWP surface coating. Microelectron. Eng. **70**(2–4), 406–411 (2003). https://doi. org/10.1016/S0167-9317(03)00286-7

[HGR06]   C.-K. Hu, L. Gignac, R. Rosenberg, Electromigration of Cu/low dielectric constant interconnects. Microelectron. Reliab. **46**(2–4), 213–231 (2006). https://doi.org/10. 1016/j.microrel.2005.05.015

[HGZ+06]   H. Haznedar, M. Gall, V. Zolotov, et al., Impact of stress-induced backflow on full-chip electromigration risk assessment. IEEE Trans. Comput. Aided Des. Integr. Circuits Syst. **25**(6), 1038–1046 (2006). https://doi.org/10.1109/tcad.2005.855941

[HLK09]   J. Hohage, M.U. Lehr, V. Kahlert, A copper-dielectric cap interface with high resistance to electromigration for high performance semiconductor devices. Microelectron. Eng. **86**(3), 408–413 (2009). https://doi.org/10.1016/j.mee.2008. 12.012

[Hoa88]   H.H. Hoang, Effects of annealing temperature on electromigration performance of multilayer metallization systems, in *Proceedings of the 26th Annual International Reliability Physics Symposium* (1988), pp. 173–178. https://doi.org/10.1109/ relphy.1988.23446

[HOG+12]   C.-K. Hu, J. Ohm, L.M. Gignac, et al., Electromigration in Cu(Al) and Cu(Mn) damascene lines. J. Appl. Phys. **111**(9), 093722–093722-6 (2012). https://doi.org/ 10.1063/1.4711070

[HRL99]   C.-K. Hu, R. Rosenberg, K.Y. Lee, Electromigration path in Cu thin-film lines. Appl. Phys. Lett. **74**(20), 2945–2947 (1999). https://doi.org/10.1063/1.123974

[HRM08]   C.S. Hau-Riege, A.P. Marathe, Z.-S. Choi, The effect of current direction on the electromigration in short-lines with reservoirs, in *IEEE International Reliability Physics Symposium (IRPS)* (2008), pp. 381–384. https://doi.org/10.1109/relphy.2008.4558916

[HRT01]   C.S. Hau-Riege, C.V. Thompson, Electromigration in Cu interconnects with very different grain structures. Appl. Phys. Lett. **78**(22), 3451–3453 (2001). https://doi.org/10.1063/1.1355304

[ITR14]   International Technology Roadmap for Semiconductors (ITRS), 2013 edn. (2014). http://www.itrs2.net/itrs-reports.html. Last retrieved on 1 Jan 2018

[ITR16]   International Technology Roadmap for Semiconductors 2.0 (ITRS 2.0), 2015 edn. (2016). http://www.itrs2.net/itrs-reports.html. Last retrieved on 1 Jan 2018

[JLH04]   H. Jiang, B. Liu, Y. Huang, et al., Thermal expansion of single wall carbon nanotubes. J. Eng. Mater. Technol. **126**, 265–270 (2004). https://doi.org/10.1115/1.1752925

[JT97]   Y.-C. Joo, C.V. Thompson, Electromigration-induced transgranular failure mechanisms in single-crystal aluminum interconnects. J. Appl. Phys. **81**(9), 6062–6072 (1997). https://doi.org/10.1063/1.364454

[KCT97]   B.D. Knowlton, J.J. Clement, C.V. Thompson, Simulation of the effects of grain structure and grain growth on electromigration and the reliability of interconnects. J. Appl. Phys. **81**(9), 6073–6080 (1997). https://doi.org/10.1063/1.364446

[KKR12]   S. Kim, D.D. Kulkarni, K. Rykaczewski, et al., Fabrication of an ultralow-resistance ohmic contact to MWCNT-metal interconnect using graphitic carbon by Electron Beam-Induced Deposition (EBID). IEEE Trans. Nanotechnol. **11**, 1223 (2012). https://doi.org/10.1109/TNANO.2012.2220377

[KRS+97]   A.B. Kahng, G. Robins, A. Singh, et al., Filling and slotting: analysis and algorithm. Tech. Rep., UCLA, Los Angeles, CA, University of Virginia, Charlottesville, VA (1997)

[KWA+13]   J. Kludt, K. Weide-Zaage, M. Ackermann, et al., Characterization of a new designed octahedron slotted metal track by simulations, in *14th International Conference on Thermal, Mechanical and Multi-Physics Simulation and Experiments in Microelectronics and Microsystems (EuroSimE)* (2013), pp. 1–5. https://doi.org/10.1109/eurosime.2013.6529907

[LDP+09]   P. Lamontagne, L. Doyen, E. Petitprez, et al., CU interconnect immortality criterion based on electromigration void growth saturation, in *IEEE International Integrated Reliability Workshop Final Report* (2009), pp. 56–59. https://doi.org/10.1109/irws.2009.5383034

[LG09]   A.R. Lavoie, F. Gstrein, Self-aligned cap and barrier. U.S. Patent App. 12/165,016 (2009)

[Lie05]   J. Lienig, Interconnect and current density stress—an introduction to electromigration-aware design, in *Proceedings of 2005 International Workshop on System Level Interconnect Prediction (SLIP)* (2005), pp. 81–88. https://doi.org/10.1145/1053355.1053374

[Lie06]   J. Lienig, Introduction to electromigration-aware physical design, in *Proceedings of the International Symposium on Physical Design (ISPD 2006)*, pp. 39–46. https://doi.org/10.1145/1123008.1123017

[LNW10]   P. Lamontagne, D. Ney, Y. Wouters, Effect of reservoir on electromigration of short interconnects, in *IEEE International Integrated Reliability Workshop Final Report (IRW)* (2010), pp. 46–50. https://doi.org/10.1109/iirw.2010.5706484

[LR71]   B.S. Landman, R.L. Russo, On a pin versus block relationship for partitions of logic graphs. IEEE Trans. Comput. **C-20**(12), 1469–1479 (1971). https://doi.org/10.1109/t-c.1971.223159

[LTC98]   W.B. Loh, M.S. Tse, L. Chan, et al., Wafer-level electromigration reliability test for deep-submicron interconnect metallization, in *Proceedings of the IEEE Hong Kong Electron Devices Meeting* (1998), pp. 157–160. https://doi.org/10.1109/hkedm. 1998.740210

[LYC+03]   J. Li, Q. Ye, A. Cassell, et al., Bottom-up approach for carbon nanotube interconnects. Appl. Phys. Lett. **82**(15), 2491–2493 (2003). https://doi.org/10. 1063/1.1566791

[LYY04]   S. Li, Z. Yu, S.-F. Yen, et al., Carbon nanotube transistor operation at 2.6 GHz. Nano Lett. **4**, 753 (2004). https://doi.org/10.1021/nl0498740

[MGL+11]   H. Mario, C.L. Gan, Y.K. Lim, et al., Effects of side reservoirs on the electromigration lifetime of copper interconnects, in *18th IEEE International Symposium on the Physical and Failure Analysis of Integrated Circuits (IPFA)* (2011), pp. 1–4. https://doi.org/10.1109/ipfa.2011.5992779

[MIH+90]   F. Matsuoka, H. Iwai, K. Hama, et al., Electromigration reliability for a tungsten-filled via hole structure. IEEE Trans. Electron Devices **37**(3), 562–568 (1990). https://doi.org/10.1109/16.47758

[MIM+07]   A. Marras, M. Impronta, I. De Munari, et al., Reliability assessment of multi-via Cu-damascene structures by wafer-level isothermal electromigration tests. Microelectron. Reliab. **47**(9–11), 1492–1496 (2007). https://doi.org/10.1016/j. microrel.2007.07.002

[MPG13]   A.M. Marconnet, M.A. Panzer, K.E. Goodson, Thermal conduction phenomena in carbon nanotubes and related nanostructured materials. Rev. Modern Phys. **85**, 1295 (2013). https://doi.org/10.1103/RevModPhys.85.1295

[MR06]   M. Meo, M. Rossi, Prediction of Young's modulus of single wall carbon nanotubes by molecular-mechanics based finite element modelling. Compos. Sci. Technol. **66** (11–12), 1597–1605 (2006). https://doi.org/10.1016/j.compscitech.2005.11.015

[MS13]   V. Mishra, S.S. Sapatnekar, The impact of electromigration in copper interconnects on power grid integrity, in *Proceedings of the Design Automation Conference (DAC)* (2013), pp. 1–6. https://doi.org/10.1145/2463209.2488842

[NSMK01]   H.V. Nguyen, C. Salm, T.J. Mouthaan, et al., Modeling of the reservoir effect on electromigration lifetime, in *Proceedings of the International Symposium on the Physical & Failure Analysis of Integrated Circuits* (2001), pp. 169–173. https:// doi.org/10.1109/ipfa.2001.941479

[OLM+01]   E. Ogawa, K.-D. Lee, H. Matsuhashi, et al., Statistics of electromigration early failures in Cu/oxide dual-damascene interconnects, in *Proceedings of the 39th Annual IEEE International Reliability Physics Symposium* (2001), pp. 341–349. https://doi.org/10.1109/relphy.2001.922925

[PAT99]   Y.-J. Park, V.K. Andleigh, C.V. Thompson, Simulations of stress evolution and the current density scaling of electromigration-induced failure times in pure and alloyed interconnects. J. Appl. Phys. **85**(7), 3546–3555 (1999). https://doi.org/10. 1063/1.369714

[PRY04]   J.-Y. Park, S. Rosenblatt, Y. Yaish, et al., Electron-phonon scattering in metallic single-walled carbon nanotubes. Nano Lett. **4**(3), 517–520 (2004). https://doi.org/ 10.1021/nl035258c

[PHR07]   M.S. Purewal, B.H. Hong, A. Ravi, et al., Scaling of resistance and electron mean free path of single-walled carbon nanotubes. Phys. Rev. Lett. **98**, 186808 (2007). https://doi.org/10.1103/PhysRevLett.98.186808

[QFX+99]   G. Qiang, L.K. Foo, Z. Xu, et al., Step like degradation profile of electromigration of W-plug contact, in *IEEE International Interconnect Technology Conference* (1999), pp. 44–46. https://doi.org/10.1109/iitc.1999.787073

[RT08]   N. Raghavan, C.M. Tan, Statistical modeling of via redundancy effects on interconnect reliability, in *15th International Symposium on the Physical and Failure Analysis of Integrated Circuits (IPFA)* (2008), pp. 1–5. https://doi.org/10.1109/ipfa.2008.4588156

[SB04]   N. Srivastava, K. Banerjee, A comparative scaling analysis of metallic and carbon nanotube interconnections for nanometer scale VLSI technologies, in *Proceedings of the 21st International VLSI Multilevel Interconnect Conference (VMIC)* (2004), pp. 393–398

[SBK05]   T. Smorodin, U. Beierlein, J.P. Kotthaus, Contacting gold nanoparticles with carbon nanotubes by self-assembly. Nanotechnology **16**(8), 1123 (2005). https://doi.org/10.1088/0957-4484/16/8/023

[Sch85]   H.-U. Schreiber, Electromigration threshold in aluminum films. Solid-State Electron. **28**(6), 617–626 (1985). https://doi.org/10.1016/0038-1101(85)90134-0

[Set09]   A. Seth, Electromigration in integrated circuits. Slides (2009). https://www.scribd.com/document/111108004/Electromigration-in-Integrated-Circuits. Last retrieved on 1 Jan 2018

[SMS+07]   W. Shao, S.G. Mhaisalkar, T. Sritharan, et al., Direct evidence of Cu/cap/liner edge being the dominant electromigration path in dual damascene Cu interconnects. Appl. Phys. Lett. **90**(5), 052106 (2007). https://doi.org/10.1063/1.2437689

[SNS+07]   D.C. Sekar, A. Naeemi, R. Sarvari, et al., Intsim: a CAD tool for optimization of multilevel interconnect networks, in *IEEE/ACM International Conference on Computer-Aided Design (ICCAD)* (2007), pp. 560–567. https://doi.org/10.1109/iccad.2007.4397324

[SSY16]   C. Subramaniam, A. Sekiguchi, T. Yamada, et al., Nano-scale, planar and multi-tiered current pathways from a carbon nanotube-copper composite with high conductivity, ampacity and stability. Nanoscale **8**, 3888 (2016). https://doi.org/10.1039/c5nr03762j

[ST10]   A. Sinitskii, J.M. Tour, Graphene electronics, unzipped. IEEE Spectr. **47**(11), 28–33 (2010). https://doi.org/10.1109/MSPEC.2010.5605889

[SYK13]   C. Subramaniam, T. Yamana, K. Kobashi, et al., One hundred fold increase in current carrying capacity in a carbon nanotube-copper composite. Nat. Commun. **4**(1), 2202 (2013). https://doi.org/10.1038/ncomms3202

[TCC96]   J. Tao, J.F. Chen, N.W. Cheung, et al., Modeling and characterization of electromigration failures under bidirectional current stress. IEEE Trans. Electron Devices **43**(5), 800–808 (1996). https://doi.org/10.1109/16.491258

[TCH93]   J. Tao, N.W. Cheung, C. Hu, Metal electromigration damage healing under bidirectional current stress. IEEE Electron Device Lett. **14**(12), 554–556 (1993). https://doi.org/10.1109/55.260787

[TF12]   C.M. Tan, C. Fu, Effectiveness of reservoir length on electromigration lifetime enhancement for ULSI interconnects with advanced technology nodes, in *11th IEEE International Conference on Solid-State and Integrated Circuit Technology (ICSICT)* (2012), pp. 1–4. https://doi.org/10.1109/icsict.2012.6467816

[Tho08]   C.V. Thompson, Using line-length effects to optimize circuit-level reliability, in *15th International Symposium on the Physical and Failure Analysis of Integrated Circuits (IPFA)* (2008), pp. 1–4. https://doi.org/10.1109/ipfa.2008.4588155

[TRO17]   A. Todri-Sanial, R. Ramos, H. Okuno, et al., A survey of carbon nanotube interconnects for energy efficient integrated circuits. IEEE Circuits Syst. Mag. **2**, 47–62 (2017). https://doi.org/10.1109/MCAS.2017.2689538

[UON+96]   M. Uekubo, T. Oku, K. Nii, et al., WN_x diffusion barriers between Si and Cu. Thin Solid Films **286**(1–2), 170–175 (1996). https://doi.org/10.1016/S0040-6090(96)08553-7

[VGH+12]   S. Van Nguyen, A. Grill, T.J. Haigh, Jr. et al., Self-aligned composite M-MOx/dielectric cap for Cu interconnect structures. U.S. Patent 8299365 (2012)

[VS81]   S. Vaidya, A.K. Sinha, Effect of texture and grain structure on electromigration in Al-0.5%Cu thin films. Thin Solid Films **75**(3), 253–259 (1981). https://doi.org/10.1016/0040-6090(81)90404-1

[WGT+08]   F.L. Wei, C.L. Gan, T.L. Tan, et al., Electromigration-induced extrusion failures in Cu/low-k interconnects. J. Appl. Phys. **104**(2), 023529–023529-10 (2008). https://doi.org/10.1063/1.2957057

[WHM+08]   F.L. Wei, C.S. Hau-Riege, A.P. Marathe, et al., Effects of active atomic sinks and reservoirs on the reliability of Cu/low-k Interconnects. J. Appl. Phys. **103**(8), 084513, 2008. https://doi.org/10.1063/1.2907962

[WVA01]   B.Q. Wei, R. Vajtai, P.M. Ajayan, Reliability and current carrying capacity of carbon nanotubes. Appl. Phys. Lett. **79**, 1172 (2001). https://doi.org/10.1063/1.1396632

[WY02]   W. Wu, J.S. Yuan, Skin effect of on-chip copper interconnects on electromigration. Solid-State Electron. **46**(12), 2269–2272 (2002). https://doi.org/10.1016/S0038-1101(02)00232-0

[XZK+12]   S. Xu, X. Zhu, H. Kotadia, et al., Remedies to control electromigration: effects of CNT doped Sn-Ag-Cu interconnects, in *62nd IEEE Electronic Components and Technology Conference (ECTC)* (2012), pp. 1899–1904. https://doi.org/10.1109/ectc.2012.6249097

[YKD00]   Z. Yao, C.L. Kane, C. Dekker, High-field electrical transport in single-wall carbon nanotubes. Phys. Rev. Lett. **84**, 2941 (2000). https://doi.org/10.1103/PhysRevLett.84.2941

[YML+11]   C.-C. Yang, F.R. McFeely, B. Li, et al., Low-temperature reflow anneals of Cu on Ru. IEEE Electron Device Lett. **32**(6), 806–808 (2011). https://doi.org/10.1109/LED.2011.2132691

[Yoo08]   C.S. Yoo, *Semiconductor Manufacturing Technology* (World Scientific, 2008). ISBN 978-981-256-823-6

[YZZ+11]   M. Yao, X. Zhang, C. Zhao, et al., Self-consistent design issues for high frequency Cu interconnect reliability incorporating skin effect. Microelectron. Reliab. **51**(5), 1003–1010 (2011). https://doi.org/10.1016/j.microrel.2010.12.011

[ZDM+00]   P. Zarkesh-Ha, J.A. Davis, J.D. Meindl, Prediction of net-length distribution for global interconnects in a heterogeneous system on-a-chip. IEEE Trans. Very Large Scale Integr. VLSI Syst. **8**(6), pp. 649–659 (2000). https://doi.org/10.1109/92.902259

Chapter 5
Summary and Outlook

This chapter summarizes the key findings and results of the book, and presents our outlook on future developments. The latter indicates how to facilitate an electromigration (EM)-compliant layout design in future EM-critical technological nodes, such as represented by the red area in Fig. 1.6 (Chap. 1).

The aim of the book has been to examine the measures available for producing an electromigration-robust layout, to compare such measures with one another and to investigate their use in practical design flows for modern integrated circuits. This approach provides circuit designers a suite of options to design and apply useful measures to prevent electromigration damage in present, and future, technology nodes. Ultimately, the challenge is to avoid exceeding permissible current densities by selectively increasing the permissible boundaries.

5.1 Summary of Electromigration-Inhibiting Measures

Semiconductor scale and thus also interconnect dimensions are continuously decreasing, due to the ongoing development of the technologies for manufacturing integrated circuits. As a result, interconnect cross-sections are being scaled back as well, which means lower boundary values for the higher current densities required for circuit functions. This trend leads to increased EM-induced problems, not only in analog circuits, but also in digital signal wires.

The *required* current density is the decisive parameter for describing the EM risk in these wires, and thus, the *permissible* current density must be increased with appropriate measures. As the current density cannot be measured at specific locations within the conductor, it must be determined with model-based measurement techniques or simulations. Other reliability indicators, apart from current density, can also be quantified in this way.

Many EM-critical parameters change as a result of decreasing semiconductor scale. For example, besides current density, mean segment lengths in interconnects

© Springer International Publishing AG 2018
J. Lienig and M. Thiele, *Fundamentals of Electromigration-Aware
Integrated Circuit Design*, https://doi.org/10.1007/978-3-319-73558-0_5

are also decreasing. They facilitate the Blech effect, which results in shorter wire segments that have increased robustness. However, the Blech effect does not go far enough to fully abate EM issues. The comparative evolutions of segment lengths and prospective Blech lengths (see the following Sect. 5.2) are not encouraging: Blech lengths are being scaled back to a greater degree than typical segment lengths. Blech lengths, in this context, are the maximum segment lengths for which no long-term damage is expected due to EM.

This means that the EM risks arising from the ever-smaller structure sizes will continue to become more prominent in the future. If we want to continue producing working circuits at such sizes, we must increase the reliability-promoting material and technology parameters. High mechanical rigidity in routing layers, a low dielectric constant for the dielectric, and a high activation energy for the interconnect material are some of the critical parameters that come under the spotlight. If these parameters cannot keep pace with requirements, reliability, usually expressed as time to failure, will suffer.

We have examined a number of different measures in this book to effectively block the effects of EM that occur with the ever-decreasing IC structure sizes. The aim of all measures and effects presented in the preceding chapters is to minimize damage occurring in a circuit's interconnect, so that the electrical parameters of the signal transmission change as little as possible within a predefined life span.

Electromigration cannot be prevented from occurring in metallic routing wires; it can only be compensated for, or restricted in its effect. This can be achieved either by reducing the wire's material transport or by raising the allowed boundary values for current density. A number of remedial measures have been proposed, which include: utilizing the bamboo effect, creating small segment lengths, producing via geometries with greater EM robustness, introducing reservoirs, using redundant vias, differentiating frequency-dependent net classes w.r.t. EM vulnerability, and selecting suitable materials. These measures are summarized below.

Bamboo effect

The bamboo effect is based on the metallic crystal lattice and the alignment of the grain boundaries perpendicular to the direction of diffusion. Diffusion typically occurs along the grain boundaries in the wire. High EM robustness can thus be achieved with small conductor cross-sections and hence fewer grain boundaries in the direction of diffusion.

Reservoirs

Reservoirs increase the maximum permissible current density by supporting the stress-migration effect, which partially neutralizes EM. Reservoirs can, however, have an adverse effect on reliability in nets with current-flow reversals, as the stress migration is reduced in this case.

Via configurations

The robustness of interconnects fabricated with dual-Damascene technology depends on whether contact is made through vias from "above" (via-above) or "below" (via-below). Via-below configurations are better from an EM-avoidance perspective than their via-above counterparts, as the higher permissible void volumes associated with via-below configurations allow higher current densities.

Redundant vias increase robustness against EM damage. They should be placed "in line" with the current direction so that all possible current paths have the same length. This ensures a uniform current distribution and, hence, avoids a local detrimental increase in current density between the vias.

Frequencies

The high frequencies typically encountered in signal nets reduce EM damage more than in power supply nets or very low-frequency nets under otherwise similar operating conditions. Hence, different current-density boundary values (limits) must be assigned to take into account signal nets and power supply nets in EM analysis.

Materials

We have examined in this book alternative materials based on carbon nanotubes (CNTs). These materials are suitable for transporting higher current densities without incurring EM damage. CNTs have a lower current-carrying capacity when integrated into a tube network or in another matrix material than when deployed on their own. A number of different CNT-based techniques, such as CNT arrays or composites, are suitable for increasing EM robustness. The choice of technique also depends on the application.

A universal scheme for preventing EM damage can only be developed by evaluating the measures mentioned above with appropriate analysis tools, such as the finite element method (FEM). Specifically, the impact of current density and other design parameters on the diffusion processes can be represented spatially by FEM, and the effects and measures analyzed by simulation. Stress migration, in particular, is an effective EM inhibitor for many measures, such as the critical length effect, reservoirs, and the type of wire contacting.

The critical length effect, described in detail in previous chapters, can be leveraged to obtain an EM-robust layout while incurring no more than a slight degrade in circuit performance. Other presented EM-inhibiting measures should be put in place that match the technology used and the interconnect parameters for the IC.

As described in detail in Chap. 4, the following practical guidelines are derived from our investigations:

- The Blech length must be considered with a view to improve time to failure.
- Reservoirs can be effectively deployed in power supply nets. However, reservoirs are not of significant value in signal nets due to the changing direction of the current, and indeed on the contrary, incorporating them could negatively impact EM robustness, depending on the manufacturing process used.
- A via-below configuration, where the vias contact the critical segment from below, should be chosen in the dual-Damascene technology if the layout permits.

- The product of length and current density must be restricted to a greater degree in the (remaining) via-above segments to compensate for the higher EM susceptibility in these segments.
- Special attention must be paid to multiple vias, as their geometrical configuration impacts the time to failure. Redundant vias are generally better than an individual via. However, possible EM benefits depend on the inter-via configuration and the vias' relationships with the connected wire segments, as higher local current densities may have an adverse effect on reliability.

State-of-the-art layout design can benefit particularly well from critical length effects. Similarly, geometries and signal frequencies can be easily considered and adjusted as necessary. By correctly tuning the technological, geometrical, and material measures described above, we estimate that the permissible current density can be increased by a factor of 10 in current semiconductor technology nodes. However, the actual reliability improvements achieved in practice will depend on numerous circuit constraints.

5.2   Outlook: Segment Lengths

As already mentioned, the comparative evolutions of segment lengths and prospective Blech lengths due to IC downscaling are not encouraging. To illustrate this, technology-dependent, EM-robust segment lengths that are achievable solely with the critical length effect (Blech length) are plotted in Fig. 5.1, based on numbers taken from the ITRS [ITR14]. The maximum current densities predicted in the roadmap are assumed for these plots.

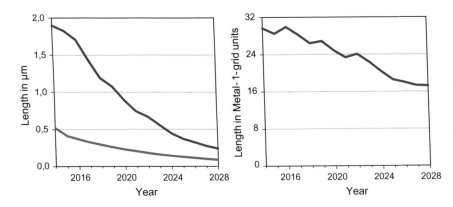

Fig. 5.1 Length limits of segments, up to which the critical length effect solely suffices for the EM robustness, depending on the respective technology node; absolute values in microns (red line on the left) and relative values in multiples of the routing grid (right). Also shown are the actual/expected mean segment lengths (blue line on the left) which drop to a lesser degree. The values are taken from ITRS [ITR14] and calculated assuming a maximum mechanical stress of 100 MPa

The red curves in Fig. 5.1 indicate that the EM-robust segment lengths are decreasing significantly as the structural miniaturization predicted in the ITRS advances [ITR16]. We can also see that these critical-length limits drop more sharply than the actual mean segment lengths on the chip (Fig. 5.1, left). This alarming realization is further supported if we plot the critical lengths w.r.t. the routing grid, i.e., in multiples of the routing grid (Fig. 5.1, right). The latter conclusion is based on the assumption that the routing grid is almost proportional to the mean segment length, as the lengths of the predominantly short segments depend on the spacing between the transistors.

Both projections imply that the number of nets that benefit from the critical length effect drops with decreasing semiconductor scale. The Blech length is exceeded in an increasing proportion of the routing—up to approximately 5% by the year 2026 (see Fig. 4.12 in Sect. 4.3.2). Remedial action, such as the introduction of reservoirs, will be required for these segments.

5.3 Outlook: Library of Electromigration-Robust Elements

It can be observed that practical tools for layout design are increasingly considering the required measures outlined in this book. However, these measures will need to be implemented as algorithms in the future, to automate the design of EM-robust integrated circuits.

One option to achieve this goal is to develop a *pattern generator* which produces routing elements for a given fabrication technology and that are EM robust when carrying a specified current density. These routing elements could be held in a library, and the routing layout then drafted exclusively with routing elements from this library (Fig. 5.2).

Consequently, the IC routing will be highly regulated, that is, *constraint-driven*, as only library elements may be used to create it. Nonetheless, the verification of EM properties is considerably simplified, because the robustness of individual elements has already been verified when the library is created. All that remains to be done in the complete layout is to examine the mutual interaction between elements when they are combined. The complexity of EM testing is thus reduced significantly, with the result that even for complex routing geometries no FE calculations are required for EM-robustness verification (Sect. 2.6.1). Furthermore, parameters can be assigned to these analyses and the results stored in the library, allowing the verification to be performed with a simplified (routing) model.

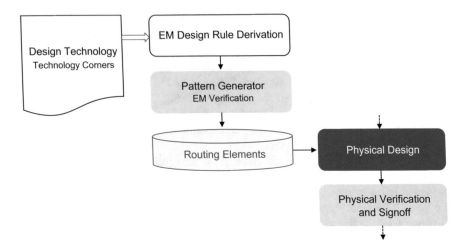

Fig. 5.2 Improving the EM robustness of the generated layout by restricting physical design to EM-robust routing elements ("layout patterns") that have been generated for a given technology and verified with special emphasis on EM properties

5.4 Outlook: New Technologies

Steps to (partially) replace copper interconnects are expected in the long term. Carbon nanotubes (CNTs) have been put forward as a viable option, as they do not suffer from EM issues; however, they are thermally destroyed at current densities greater than 10^9 A/cm^2. There is a range of possible applications for single CNTs, CNT arrays, and variously filled composites. These options must be evaluated for the prospective applications, such as interconnects, vias, and contact structures.

The high current-carrying capacity of CNTs has been verified by many research groups (Sect. 4.9). However, current-density values that are viable for use in practice depend on the implemented, usable technology. Realistic current densities for composites, for example, are determined by the matrix material, the lengths, and the orientation of the CNTs. EM can still occur in a copper-CNT composite, where some current flows through the metal. By contrast, the minimum electrical resistance is restricted if a polymer is used. The maximum current density is lower as well with CNT arrays due to mutual interactions.

All of these issues mentioned above must be overcome before routing systems for integrated circuits can be built with CNTs. We anticipate that a number of new routing constraints will emerge as the layout design process is adapted to these technologies.

5.5   Electromigration-Aware Design: Driven by Constraints

New challenges for IC design are appearing on the back of the ongoing trend in IC downscaling, i.e., structural miniaturization. Physical designs with ever-smaller feature sizes are subjected to an increasing number of more complex constraints due to these challenges. They are increasingly curtailing freedom in the design flow and are setting the boundaries of the ever-decreasing solution space. Hence, we are witnessing a slow, but steady evolution from a *constraint-correct* design flow to a *constraint-driven* one. With the latter, design algorithms and methodologies not only verify the correct implementation of constraints, but rather are governed by them.

One such challenge is EM. EM considerations are thus producing additional constraints in the design flow that are becoming the principal obstacles arising from progressive reductions in structure size. The resultant reduction in the available solution space for IC and routing design is illustrated in Fig. 5.3. Hence, a distinction must be made in the future between EM-robust and non-viable routing elements, whereby only EM-robust elements may be used for routing. Thus, constraint-driven routing is expected to predominate in future.

This book has examined the underlying EM-inhibiting effects and proposed guidelines for creating routing elements that are more EM robust. The prospective rules for the constraint-driven design in general and, in particular, constraint-driven routing techniques can be derived from the presented guidelines.

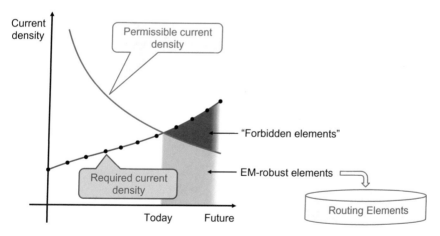

Fig. 5.3 Projected evolution of the routing solution space with falling current-density boundaries (green) and increasing required current densities (red, cf. Fig. 1.6 in Chap. 1). The solution space for the allowed routing elements will be increasingly curtailed; hence, today's constraint-correct routing evolves into constraint-driven routing where only EM-robust elements may be used (see Fig. 5.2 for the generation of the routing elements)

We expect that future, nanoscale design of reliable integrated circuits can only be achieved by applying the methodologies summarized in this final chapter and described in detail throughout this book.

References

[ITR14]   Int. Technology Roadmap for Semiconductors (ITRS), 2013 edn (2014). http://www. itrs2.net/itrs-reports.html. Last retrieved on 1 Jan 2018
[ITR16]   Int. Technology Roadmap for Semiconductors 2.0 (ITRS 2.0), 2015 edn (2016). http:// www.itrs2.net/itrs-reports.html. Last retrieved on 1 Jan 2018

Index

© Springer International Publishing AG 2018
J. Lienig and M. Thiele, *Fundamentals of Electromigration-Aware Integrated Circuit Design*, https://doi.org/10.1007/978-3-319-73558-0

Printed in the United States
By Bookmasters